WAVE PROPAGATION IN DISSIPATIVE MATERIALS

A REPRINT OF FIVE MEMOIRS

BY

B. D. COLEMAN, M. E. GURTIN, I. HERRERA R., AND C. TRUESDELL

SPRINGER-VERLAG · NEW YORK INC. 1965

ISBN 978-3-642-88693-5 ISBN 978-3-642-88691-1 (eBook)
DOI 10.1007/978-3-642-88691-1

Softcover reprint of the hardcover 1st edition 1965

Library of Congress Catalog Card Number 65-25011

Title No. 1306

Preface

Common experience reveals two basic aspects of wave propagation. First, while preserving their identity and travelling at definite speeds, sounds finally die out. Second, weak sounds may combine to form strong noises. Theories of acoustic propagation have succeeded in representing these aspects of experience separately, but never combined as in nature. The classical theories of sound in perfect fluids and elastic solids easily yield common speeds of propagation for plane infinitesimal disturbances, but no damping. Moreover, within EULER's theory of the perfect fluid, or its generalization, the GREEN-KIRCHHOFF-KELVIN theory of finite elasticity, weak waves may grow stronger and become shock waves, which propagate according to more complicated but equally definite principles. Effects of internal damping are easily added for theories of infinitesimal deformation, but for finite motions a dead end was reached about sixty years ago.

Indeed, in 1901 DUHEM proved that according to the NAVIER-STOKES theory of fluids acceleration waves and waves of higher order cannot exist, and for shock waves he claimed a similar result, which has since been shown to be valid subject to certain qualifications. So as to save the phenomena of sound and noise, as was necessary if the NAVIER-STOKES theory was to deserve the place proposed for it as a refinement upon EULER's theory, DUHEM introduced the concept of "quasi-wave", a region of rapid but continuous transition. In 1906 PRANDTL offered the same argument in a more special and less precise way, and from his remarks grew, eventually, not only the theory of the plane "shock layer" with its notorious mathematical difficulties, but also the widespread opinion that any kind of "dissipative mechanism" smoothes out discontinuities.

In 1930 LAMPARIELLO observed that this opinion is false. Indeed, the linear hyperbolic partial differential equation

$$T\frac{\partial^2 u}{\partial x^2}+F\frac{\partial u}{\partial t}=\sigma\frac{\partial^2 u}{\partial t^2},$$

which represents the small transverse motion of a perfectly flexible string with a frictional resistance proportional to the velocity, clearly *admits discontinuous solutions of all orders*. Moreover, their speed of propagation is $U=\sqrt{T/\sigma}$, unaffected by the magnitude of the coefficient of friction F. The idea is easily generalized, and in the vast Italian literature on accumulative theories ("fisica ereditaria"), little or no attention has been paid to waves, since in the equations considered there certain accumulative terms ("termini ereditari") involving only derivatives of order lower than that of the differential system are added to account for dissipation. Of course, with shock waves the argument no longer applies, but the theories are linearized from the start, with only infinitesimal deformations in view, so the question need not arise.

The non-Italian literature, meanwhile, continued to regard surfaces of discontinuity incompatible with "dissipative mechanisms" until just the last few years, when solutions involving shock waves according to BOLTZMANN'S theory of infinitesimal visco-elasticity began to be noticed and discussed.

The scene was now set for the four remarkable memoirs reprinted in this volume. These provide the first theory of waves that accounts for *all common acoustic experience.*

Essential to the achievement here recorded was the general development of continuum mechanics in the past twenty years, and in particular the following three mathematical structures. (1) The theory of simple materials as formulated by NOLL in 1957—8. (2) The theory of fading memory as defined and applied by COLEMAN & NOLL in 1959—60. (3) The thermodynamics of simple materials as created by COLEMAN in 1964. Very helpful also was the development of technique in the theory of singular surfaces in non-dissipative media by T. Y. THOMAS in the late forties and early fifties, followed by ERICKSEN'S definitive memoir of 1953 on waves in incompressible isotropic elastic materials.

While this whole complex of ideas and the associated mathematics are used here, the key to resolution of the apparent paradox of wave propagation in dissipative materials may be found in the fact that although the NAVIER-STOKES theory emerges according to COLEMAN & NOLL'S scheme as an *asymptotic approximation* in the limit of slow motion for any simple fluid with fading memory, it is itself an exceptional material in that it does *not* exhibit fading memory in COLEMAN & NOLL'S sense. COLEMAN, GURTIN & HERRERA have proved that the property of fading memory *removes the inconsistency between sharp discontinuities and dissipation*, so that, in accord with experience, *both damping and propagation* become possible in theories of materials. The inability of the NAVIER-STOKES fluid to support wave motions merely illustrates its exceptional character, untypical of real fluid behavior. Looking backward now, we can say that the theory of linearly viscous fluids, although properly invariant, was misleading as a basis for the kind of ritual which is called "intuition" by those who consider themselves innately endowed with knowledge of physics. BOLTZMANN'S theory of infinitesimal visco-elastic materials is closer to nature in regard to wave propagation, yet in turn untypical because of the linearization, which renders all kinds of waves equivalent and removes the possibility of reinforcement. The theory of COLEMAN, GURTIN, & HERRERA unifies the HUGONIOT-HADAMARD theory of weak and strong waves in finitely deformed elastic fluids and solids with the recent studies of waves in infinitesimal visco-elasticity and extrapolates between them as limit cases; it takes account both of finite deformations and of long-range memory.

There is not space here to summarize the papers reprinted in this volume. I confine myself to pointing out two remarkable results. First, for a given material there exists a critical jump in the density rate [Remark 5.2 in Part II (p. 255)]: If the instantaneous modulus of compression is positive, a plane compressive acceleration wave carrying a lesser jump is damped out steadily as it progresses into a region at rest, but the amplitude of a wave carrying a greater jump becomes infinite within a finite time. In Part III this result is shown to remain valid when thermodynamic influences are taken into account. Thus appears, for the first time in a mathematical theory, an explanation for the damping of sufficiently weak sounds and the explosion of sufficiently strong ones. Second, in Part IV thermodynamic limitations are used to show that the acoustic tensor in a non-conductor or a definite conductor of heat is a symmetric tensor, a result which suggests an experimental test of the theory here developed. I remark also that the general theory presented here puts singular surfaces back in place as *the best model* for propagation of nearly all kinds of sharp disturbances, except, of course, in cases where the finite thickness of a region of transition is really of prime interest.

When I proposed the immediate reprinting of these four memoirs on the general theory, Messrs. COLEMAN and GURTIN asked me to include my paper of 1961 on waves in elastic materials. While its subject is strictly excluded by the title of the volume, it may be useful as a reference, since in many cases the theorems on waves in dissipative materials are phrased by reduction to the theory of waves in a particular purely elastic material, determined by the constitutive functional and the deformation-temperature history together. For the reprinting, a few misprints and slips have been corrected, and a few remarks, set off in braces, have been added. May the inclusion of my paper, with its dedication and references, remind the reader of the great tradition of HUGONIOT, HADAMARD, and DUHEM, here vindicated and refurbished.

This preface must close by an expression of gratitude to Springer-Verlag, not only for its peerless typography and accuracy, but also for the elegant courtesy, vanishing relic of old times, which it extends to its authors and editors.

C. TRUESDELL

Contents

I. The Velocity of One-Dimensional Shock and Acceleration Waves

BERNARD D. COLEMAN, MORTON E. GURTIN & ISMAEL HERRERA R.

Contents

1. Introduction

The stress at a material point of a simple material [*1*] depends on the entire temporal history of the strain at the point. The functional expressing this dependence is given certain smoothness properties by the principle of fading memory [*2, 3, 4*]. We here discuss the motion of singularities in the configuration of a general simple material. Our interest is in shock and acceleration waves. Shocks involve jump discontinuities in strain and velocity, while acceleration waves involve jump discontinuities in the gradient of the strain and the acceleration.

To focus attention on the *smoothness* properties of the stress functional which govern the propagation of discontinuities, we attempt to avoid geometrical complications by offering a one-dimensional theory. We consider only the motion of those disturbances which can be described with a single spatial coordinate. In Section 6 we give examples of special situations in which the behavior of a three-dimensional body may be described by our one-dimensional theory. Until that section all the kinematical and mechanical concepts we use are one-dimensional; the strain γ and the stress T (or tension) are here not second-order tensors, but just numbers. We let a closed interval of the real line describe a reference configuration $\mathscr{R}$ of the body in question. A motion of the body is then a function $x = x(X, \tau)$ giving the location x at time τ of the material point which has the position X in the reference configuration. We often identify a material point with its position X in the reference configuration. When the derivatives

$$F(X,\tau) = \frac{\partial}{\partial X} x(X,\tau), \qquad \dot{x}(X,\tau) = \frac{\partial}{\partial \tau} x(X,\tau), \qquad \ddot{x}(X,\tau) = \frac{\partial^2}{\partial \tau^2} x(X,\tau) \qquad (1.1)$$

exist, we call them, respectively, the *deformation gradient*, the *velocity*, and the *acceleration* of X at time τ. The *strain* γ is then

$$\gamma = F - 1. \tag{1.2}$$

Although our terminology may seem to imply that we have in mind longitudinal motion of an isotropic cylinder,* we indicate in Section 6 that with minor modifications our theory may be applied also to pure shear waves** in incompressible isotropic media.

The *linear theory of viscoelasticity* is based on the constitutive assumption

$$T(X,t) = T_0 + G(0)\,\gamma(X,t) + \int_0^\infty \frac{dG(s)}{ds}\,\gamma(X,t-s)\,ds. \tag{1.3}$$

Here $T(X,t)$ is the stress at X at time t, $\gamma(X,t-s)$ is the strain at X at time $t-s$, $0 \leqq s < \infty$, T_0 is the residual stress, *i.e.*, the stress the material exhibits when left at equilibrium in the reference configuration, and $G(\cdot)$ is a material function which we call the *stress-relaxation function*. It is clear from (1.3) that $G(0)$ governs the instantaneous response to changes in strain. Intuition suggests that in linear viscoelasticity the velocity of propagation of both shocks and acceleration waves should be given by a single formula completely analogous to that familiar in linear elasticity, with the initial value $G(0)$ of the stress relaxation function playing the role of the elastic modulus. A formal proof of this assertion has been given by Fisher & Gurtin [*5*] and Herrera & Gurtin [*6*].***

In a general *simple material*, instead of (1.3), we have

$$T(X,t) = \mathop{\mathfrak{T}}_{s=0}^{\infty}\left(F^{(t)}(X,s)\right), \tag{1.4}$$

where

$$F^{(t)}(X,s) = F(X,t-s), \qquad 0 \leqq s < \infty, \tag{1.5}$$

is the deformation gradient at time $t-s$ and $\mathfrak{T}$ is a *functional* (which certainly need not be linear) of the *history* $F^{(t)}(X,\cdot)$. Although $F^{(t)}(X,\cdot)$ and $T(X,t)$ depend on the material point X, the form of the functional $\mathfrak{T}$ in (1.4) is assumed independent of X; this means that we assume the medium to be intrinsically homogeneous and choose a homogeneous configuration $\mathscr{R}$ as a reference.

Let us write (1.4) and (1.5) without making explicit the understood dependence of T and F on X:

$$T(t) = \mathop{\mathfrak{T}}_{s=0}^{\infty}\left(F^{(t)}(s)\right), \tag{1.6}$$

$$F^{(t)}(s) = F(t-s), \qquad 0 \leqq s < \infty. \tag{1.7}$$

A knowledge of the function $F^{(t)}(\cdot)$ of (1.7), whose domain of definition is $[0, \infty)$, is equivalent to a knowledge of both the *present value* $F(t)$ and the *past history*

---

* *I.e.*, plane dilatational waves.

** *I.e.*, plane transverse waves.

*** Shock waves are discussed in [*5*] and acceleration waves in [*6*].

$F_{(r)}^{(t)}(\cdot)$, where $F_{(r)}^{(t)}(\cdot)$ is the restriction of $F^{(t)}(\cdot)$ to the open interval $(0, \infty)$.* Hence (1.6) can be written in the form [8]

$$T(t) = \underset{s=0}{\overset{\infty}{\mathfrak{T}}}\left(F_{(r)}^{(t)}(s); F(t)\right). \tag{1.8}$$

For each value of $F(t)$, $\mathfrak{T}$ in (1.8) is a functional of $F_{(r)}^{(t)}(\cdot)$, while for each past history $F_{(r)}^{(t)}(\cdot)$, $\mathfrak{T}$ is an "ordinary function" of $F(t)$. The smoothness properties given to $\mathfrak{T}$ by the principle of fading memory guarantee the existence of a functional $D\mathfrak{T}$ defined by**

$$D\underset{s=0}{\overset{\infty}{\mathfrak{T}}}\left(F^{(t)}(s)\right) = \frac{\partial}{\partial F(t)}\underset{s=0}{\overset{\infty}{\mathfrak{T}}}\left(F_{(r)}^{(t)}(s); F(t)\right). \tag{1.9}$$

Such "D operators" were introduced by COLEMAN [7, 8], who showed that they play a central role in the thermodynamics of simple materials. Since the partial derivative in (1.9) is taken holding the past history fixed, the "modulus"

$$E = D\underset{s=0}{\overset{\infty}{\mathfrak{T}}}\left(F^{(t)}(s)\right) \tag{1.10}$$

determines the instantaneous response to a small strain impulse superposed on $F^{(t)}(\cdot)$ at time t. For this reason we call E the *instantaneous tangent modulus* corresponding to the history $F^{(t)}(\cdot)$. Note that E is not a material constant: It depends on the history $F^{(t)}(\cdot)$. The *instantaneous secant modulus* E_Λ corresponding to the history $F^{(t)}(\cdot)$ and the jump Λ is

$$E_\Lambda = \frac{1}{\Lambda}\left[\underset{s=0}{\overset{\infty}{\mathfrak{T}}}\left(F_{(r)}^{(t)}(s); F(t)+\Lambda\right) - \underset{s=0}{\overset{\infty}{\mathfrak{T}}}\left(F_{(r)}^{(t)}(s); F(t)\right)\right]; \tag{1.11}$$

of course,

$$\lim_{\Lambda\to 0} E_\Lambda = E. \tag{1.12}$$

For a material obeying (1.3), $E = E_\Lambda = G(0)$. In general, however, $E \neq E_\Lambda$.

For a perfectly elastic material in finite strain, E_Λ and E reduce to the usual secant and tangent moduli of elasticity. It is known that in such a material the intrinsic velocity of propagation*** U of a shock is given by

$$U = \sqrt{\frac{E_\Lambda}{\varrho_0}}, \tag{1.13}$$

where Λ is the jump in the deformation gradient across the shock and ϱ_0 is the mass density in the reference configuration. For an acceleration wave in an elastic material

$$U = \sqrt{\frac{E}{\varrho_0}}. \tag{1.14}$$

* We do not always assume that $F^{(t)}(s)$ is continuous at $s=0$.

** This is made precise in Section 2. For a more detailed discussion see [8].

*** We give a precise definition of U in Section 3. To make the formulae (1.13), (1.14) appear more familiar to some readers, we here remark that $U = F_1^{-1}(u - \dot{x}_1) = F_2^{-1}(u - \dot{x}_2)$ where u is the wave velocity, *i.e.*, the ordinary velocity of the wave as seen by an observer at rest; F_1 and $\dot{x}_1$ are the deformation gradient and velocity of the medium just ahead of the wave, while F_2 and $\dot{x}_2$ are the corresponding quantities just behind the wave front. The continuity across an acceleration wave is such that $\dot{x}_1 = \dot{x}_2$ and $F_1 = F_2$, but for a shock we have not only $\dot{x}_1 \neq \dot{x}_2$ but also $F_1 - F_2 = \Lambda \neq 0$.

We here show that when E_A is defined by (1.11) and E by (1.12), the equations (1.13) and (1.14) hold for *all* simple materials obeying the principle of fading memory. Thus we extend to general simple materials formulae for wave speeds known previously only for perfectly elastic and linearly viscoelastic materials.

We state now, once and for all, that the expressions we derive here for the velocity of propagation of singularities give only *necessary* conditions for the existence of singularities. We make no attempt to exhibit explicit solutions of the dynamical equations showing the discontinuities we discuss.

This paper ignores thermodynamics. In future articles for this Archive, COLEMAN & GURTIN will discuss the growth and decay of acceleration waves and the restrictions placed by the second law on the motion of shock and acceleration waves in media with memory.

2. The Constitutive Functional

General Smoothness Properties

Let $\mathscr{H}$ be an arbitrary normed linear manifold of the space of real functions on $(0, \infty)$. For the norm of $\mathscr{H}$ we write $\|\cdot\|$. By $\mathscr{H}^{\oplus}$ we always mean the set of real functions $f(\cdot)$ defined on $[0, \infty)$ and such that the restriction $f_{(r)}(\cdot)$ of $f(\cdot)$ to the open interval $(0, \infty)$ belongs to $\mathscr{H}$. Clearly $\mathscr{H}^{\oplus}$ is a linear space. We define the norm of a function $f(\cdot)$ in $\mathscr{H}^{\oplus}$ by*

$$\|f(\cdot)\|^{\oplus} = \|f_{(r)}(\cdot)\| + |f(0)|. \tag{2.1}$$

Without specifying further either $\mathscr{H}$ or its norm $\|\cdot\|$, we now state our

Smoothness hypothesis on the functional $\mathfrak{T}(\cdot)$ *of* (1.6): We suppose that $\mathfrak{T}$ is defined on each function $f(\cdot)$ in $\mathscr{H}^{\oplus}$ and has a functional derivative in the sense of Fréchet:

$$\underset{s=0}{\overset{\infty}{\mathfrak{T}}}(f(s)+g(s)) = \underset{s=0}{\overset{\infty}{\mathfrak{T}}}(f(s)) + \delta^{\oplus}\underset{s=0}{\overset{\infty}{\mathfrak{T}}}(f(s)\,|\,g(s)) + o\left(\|g(\cdot)\|^{\oplus}\right). \tag{2.2}$$

The Fréchet derivative $\delta^{\oplus}\underset{s=0}{\overset{\infty}{\mathfrak{T}}}(f(s)\,|\,g(s))$ is here assumed to be not only linear and continuous in $g(\cdot)$, but also jointly continuous in $f(\cdot)$ and $g(\cdot)$.

When using the notation of (1.8) we put**

$$\underset{s=0}{\overset{\infty}{\mathfrak{T}}}(k(s);\beta) = \underset{s=0}{\overset{\infty}{\mathfrak{T}}}(f(s)), \tag{2.3}$$

where

$$k(\cdot) = f_{(r)}(\cdot), \qquad \beta = f(0). \tag{2.4}$$

* The idea that equations (2.1) and (2.2), with $\|\cdot\|$ replaced by $\|\cdot\|_h$, supply a simple and direct statement of the principle of fading memory as used in [7] and [8] first arose in May 1964 in conversations between L. E. BRAGG and B. D. COLEMAN.

** A purist would use different symbols for the operators "$\mathfrak{T}$" on the two sides of (2.3). We, however, rely on the presence or absence of the semicolon to indicate which point of view we take. When the independent variables are not shown, if we do not wish to emphasize the distinction we write simply $\mathfrak{T}$, but if the distinction is important, we write $\mathfrak{T}(\cdot)$ or $\mathfrak{T}(\cdot;\cdot)$. Note that we use $\delta^{\oplus}\mathfrak{T}$ to denote the "total" Fréchet derivative of $\mathfrak{T}$, *i.e.*, the Fréchet derivative of $\mathfrak{T}(\cdot)$, while $\delta\mathfrak{T}$ denotes the "partial" Fréchet derivative of $\mathfrak{T}(\cdot;\cdot)$ with respect to its first variable.

Clearly, our smoothness hypothesis on $\mathfrak{T}(\cdot)$ implies the following *smoothness properties for* $\mathfrak{T}(\cdot;\cdot)$: For each number β, $\underset{s=0}{\overset{\infty}{\mathfrak{T}}}(k(s),\beta)$ is defined for every function $k(\cdot)$ in $\mathscr{H}$ and has a partial Fréchet derivative with respect to $k(\cdot)$,

$$\underset{s=0}{\overset{\infty}{\mathfrak{T}}}(k(s)+w(s);\beta)=\underset{s=0}{\overset{\infty}{\mathfrak{T}}}(k(s);\beta)+\delta\underset{s=0}{\overset{\infty}{\mathfrak{T}}}(k(s);\beta\,|\,w(s))+o(\|w(\cdot)\|); \tag{2.5}$$

this Fréchet derivative $\delta\underset{s=0}{\overset{\infty}{\mathfrak{T}}}(k(s);\beta\,|\,w(s))$ is jointly continuous in $k(\cdot)$, β, and $w(\cdot)$. Further, for each fixed function $k(\cdot)$ in $\mathscr{H}$, $\underset{s=0}{\overset{\infty}{\mathfrak{T}}}(k(s);\beta)$ has a partial derivative with respect to β,

$$\frac{\partial}{\partial\beta}\underset{s=0}{\overset{\infty}{\mathfrak{T}}}(k(s);\beta)=D\underset{s=0}{\overset{\infty}{\mathfrak{T}}}(k(s);\beta), \tag{2.6}$$

that is jointly continuous in $k(\cdot)$ and β.

The results (2.5) and (2.6) are immediate consequences of the identity

$$\delta^{\oplus}\underset{s=0}{\overset{\infty}{\mathfrak{T}}}(f(s)\,|\,g(s))=\delta\underset{s=0}{\overset{\infty}{\mathfrak{T}}}(f_{(r)}(s);f(0)\,|\,g_{(r)}(s))+D\underset{s=0}{\overset{\infty}{\mathfrak{T}}}(f_{(r)}(s);f(0))\,g(0). \tag{2.7}$$

Let $F^{(t)}(\cdot)$ be an arbitrary history of the deformation gradient with the property that $F^{(t)}_{(r)}(\cdot)$, the restriction of $F^{(t)}(\cdot)$ to $(0,\infty)$, is in $\mathscr{H}$. Let $\tilde{F}^{(t)}(\cdot)$ be the history obtained from $F^{(t)}(\cdot)$ by superimposing a strain impulse with jump Λ at time t; *i.e.*,

$$\tilde{F}^{(t)}(s)=\begin{cases}F^{(t)}(0)+\Lambda, & s=0,\\ F^{(t)}(s), & 0<s<\infty.\end{cases} \tag{2.8}$$

Note that the past histories $F^{(t)}_{(r)}(\cdot)$ and $\tilde{F}^{(t)}_{(r)}(\cdot)$ corresponding to $F^{(t)}(\cdot)$ and $\tilde{F}^{(t)}(\cdot)$ are the same, but the corresponding present values $F(t)$, $\tilde{F}(t)$ of the deformation gradient differ,

$$F(t)=F^{(t)}(0),\qquad \tilde{F}(t)=\tilde{F}^{(t)}(0)=F(t)+\Lambda, \tag{2.9}$$

and so do the present values $T(t)$, $\tilde{T}(t)$ of the stress,

$$\begin{aligned}T(t)&=\underset{s=0}{\overset{\infty}{\mathfrak{T}}}(F^{(t)}_{(r)}(s);F(t)),\\ \tilde{T}(t)&=\underset{s=0}{\overset{\infty}{\mathfrak{T}}}(\tilde{F}^{(t)}_{(r)}(s);\tilde{F}(t))=\underset{s=0}{\overset{\infty}{\mathfrak{T}}}(F^{(t)}_{(r)}(s);F(t)+\Lambda).\end{aligned} \tag{2.10}$$

In the notation of (2.10), the definition (1.11) becomes

$$E_\Lambda=\frac{\tilde{T}(t)-T(t)}{\Lambda}, \tag{2.11}$$

which motivates calling E_Λ the *instantaneous secant modulus for the history* $F^{(t)}(\cdot)$ *and the jump* Λ. Furthermore, (1.12) may be written

$$E=\lim_{\Lambda\to 0}\frac{\tilde{T}(t)-T(t)}{\Lambda}, \tag{2.12}$$

which is why we call E the *instantaneous tangent modulus*.

The Theory of Fading Memory

Our main results are sufficiently general that neither $\|\cdot\|$ nor $\mathscr{H}$ need be specified. In some applications, however, $\mathscr{H}$ may be thought of as the Hilbert space $\mathscr{H}_h$ of the theory of fading memory [2, 3, 4, 7]. This space, with its norm $\|\cdot\|_h$, may be defined as follows. Let h be a fixed influence function, by which we mean a positive, monotone-decreasing, continuous function on $[0, \infty)$, decaying fast enough to be square integrable. We put

$$\|f(\cdot)\|_h = \sqrt{\int_0^\infty f(s)^2 h(s)^2 ds} \tag{2.13 a}$$

and let $\mathscr{H}_h$ be the set of all real functions $f(\cdot)$ on $(0, \infty)$ such that $\|f(\cdot)\|_h < \infty$. The inner product $\langle\cdot, \cdot\rangle_h$ corresponding to the norm (6.1) is

$$\langle f(\cdot), g(\cdot)\rangle_h = \int_0^\infty f(s)\, g(s)\, h(s)^2 ds. \tag{2.13 b}$$

Relation to Linear Viscoelasticity

To see how the theory of linear viscoelasticity falls as a special case within our present framework, we put $\gamma = F - 1$ in (1.3) and obtain

$$T(t) = T_0 + G(0)\, F(t) + \int_0^\infty \frac{dG(s)}{ds} F(t-s)\, ds - G(\infty). \tag{2.14}$$

Thus in linear viscoelasticity one is taking the following special choices for $\mathfrak{T}(\cdot)$ and $\mathfrak{T}(\cdot;\cdot)$:

$$\overset{\infty}{\underset{s=0}{\mathfrak{T}}}\big(f(s)\big) = T_0 + G(0)\, f(0) + \int_0^\infty \frac{dG(s)}{ds} f(s)\, ds - G(\infty), \tag{2.15 a}$$

$$\overset{\infty}{\underset{s=0}{\mathfrak{T}}}\big(k(s);\beta\big) = T_0 + G(0)\,\beta + \int_0^\infty \frac{dG(s)}{ds} k(s)\, ds - G(\infty). \tag{2.15 b}$$

It is easily seen that if we choose $\mathscr{H} = \mathscr{H}_h$, $\|\cdot\| = \|\cdot\|_h$, and assume that $h(s)^{-1}\frac{dG(s)}{ds}$ is square integrable, then the functional $\mathfrak{T}$ in (2.15 a) satisfies our smoothness hypothesis. Furthermore (2.15 b) yields

$$D\overset{\infty}{\underset{s=0}{\mathfrak{T}}}\big(k(s);\beta\big) = G(0), \tag{2.16a}$$

$$\delta\overset{\infty}{\underset{s=0}{\mathfrak{T}}}\big(k(s);\beta\,|\,w(s)\big) = \int_0^\infty \frac{dG(s)}{ds} w(s)\, ds. \tag{2.16b}$$

The observation that $D\mathfrak{T}$ and $\delta\mathfrak{T}$ are here independent of $k(\cdot)$ and β is, of course, a consequence of the linearity of the special functional (2.15 b). We note that it follows from (1.10), (1.11), (2.15 b) and (2.16 a) that in linear viscoelasticity

$$E_A = E = G(0). \tag{2.17}$$

When the principle of fading memory is assumed, *i.e.*, when $\mathcal{H}=\mathcal{H}_h$ and $\|\cdot\|=\|\cdot\|_h$, the theory of linear viscoelasticity is not only a special case of our present theory, but it also has the property that every functional in our theory is *approximated* by some functional of linear viscoelasticity in the limit of *small deformations*. Following arguments given by COLEMAN & NOLL [3], we now make this idea precise. Let $1^\dagger(\cdot)$ and $1^\dagger_{(r)}(\cdot)$ be the (constant) functions on $(0,\infty)$ and $[0,\infty)$ whose values are always unity:

$$\begin{aligned} 1^\dagger(s) &= 1, \qquad 0\leqq s<\infty \\ 1^\dagger_{(r)}(s) &= 1, \qquad 0<s<\infty. \end{aligned} \tag{2.18}$$

If a body is held at rest in its reference configuration $\mathcal{R}$ at all times, then $F^{(t)}(\cdot)\equiv 1^\dagger(\cdot)$, i.e., $F(t)=1$ and $F^{(t)}_{(r)}(\cdot)=1^\dagger_{(r)}(\cdot)$. Hence

$$T_o = \mathop{\mathfrak{T}}_{s=0}^{\infty}\left(1^\dagger(s)\right) \tag{2.19}$$

is the residual stress corresponding to perpetual rest in the reference configuration. It follows from our smoothness hypothesis (2.2) that when $\|F^{(t)}(\cdot)-1^\dagger(\cdot)\|^{\oplus}$ is small, the difference between $T(t)$ and T_o is approximately linear in $F^{(t)}(\cdot)-1^\dagger(\cdot)$:

$$T(t) = T_o + \delta^{\oplus} \mathop{\mathfrak{T}}_{s=0}^{\infty}\left(1^\dagger(s)\,|\,F^{(t)}(s)-1^\dagger(s)\right) + o\left(\|F^{(t)}(s)-1^\dagger(s)\|^{\oplus}\right). \tag{2.20}$$

Let us put

$$\gamma(t)=F(t)-1, \qquad \gamma^{(t)}(s)=\gamma(t-s)=F(t-s)-1, \qquad 0<s<\infty. \tag{2.21}$$

Note that we take the domain of $\gamma^{(t)}(\cdot)$ to be $(0,\infty)$. Clearly, we can write (2.20) in the form

$$\begin{aligned} T(t) = T_o &+ \delta \mathop{\mathfrak{T}}_{s=0}^{\infty}\left(1^\dagger_{(r)}(s);1\,|\,\gamma^{(t)}(s)\right) + D \mathop{\mathfrak{T}}_{s=0}^{\infty}\left(1^\dagger_{(r)}(s);1\right)\gamma(t) + \\ &+ o\left(\|\gamma^{(t)}(\cdot)\| + |\gamma(t)|\right). \end{aligned} \tag{2.22}$$

Equation (2.21) holds for every choice of $\mathcal{H}$. We suppose now that $\mathcal{H}=\mathcal{H}_h$. Then, since $\delta \mathop{\mathfrak{T}}_{s=0}^{\infty}\left(1^\dagger_{(r)}(s);1\,|\,\cdot\right)$ is a continuous linear functional over a Hilbert space, it can be represented as an inner product; *i.e.*, there exists a function $K(\cdot)$ in $\mathcal{H}_h$ such that for all $f(\cdot)$ in $\mathcal{H}_h$

$$\delta \mathop{\mathfrak{T}}_{s=0}^{\infty}\left(1^\dagger_{(r)}(s);1\,|\,f(s)\right) = \int_0^\infty K(s)\,f(s)\,h(s)^2\,ds. \tag{2.23}$$

Letting $G(\cdot)$ be the unique solution of

$$\frac{d}{ds}G(s) = K(s)\,h(s)^2, \qquad G(0) = D \mathop{\mathfrak{T}}_{s=0}^{\infty}\left(1^\dagger_{(r)}(s);1\right), \tag{2.24}$$

we can write (2.22) in the form

$$T(t) = T_o + \int_0^\infty \frac{dG(s)}{ds}\,\gamma^{(t)}(s)\,ds + G(0)\,\gamma(t) + o\left(\|\gamma^{(t)}(\cdot)\|_h + |\gamma(t)|\right). \tag{2.25}$$

Let us put

$$\varepsilon = \sup_{-\infty<\tau<\infty} |F(\tau) - 1| \tag{2.26}$$

and suppose that ε is small; *i.e.*, that the absolute value of the strain γ is small at all times. Clearly $|\gamma(\tau)| = O(\varepsilon)$, and equation (2.13a) implies that

$$\|\gamma^{(t)}(\cdot)\|_h = O(\varepsilon). \tag{2.27}$$

Hence the error term in (2.25) is $o(\varepsilon)$:

$$o\left(\|\gamma^{(t)}(\cdot)\|_h + |\gamma(t)|\right) = o(\varepsilon). \tag{2.28}$$

We have justified the following

Remark. If $\mathscr{H} = \mathscr{H}_h$ and $\|\cdot\| = \|\cdot\|_h$ in our smoothness hypotheses on $\mathfrak{T}$, and if ε is defined by (2.26), then to within an error $o(\varepsilon)$ the stress is given by the constitutive equation (1.3) of linear viscoelasticity with T_0 defined by (2.19) and $G(\cdot)$ by (2.24).

3. General Properties of Waves

Let $x = x(X, \tau)$ be the position at time τ of the material point whose position in the reference configuration $\mathscr{R}$ is X. The *material representation* of a wave* is a smooth one-parameter family of points $Y(\tau) \in \mathscr{R}$, $-\infty < \tau < \infty$, such that $Y(\tau)$ gives the material point (labeled by its position in $\mathscr{R}$) at which the wave is to be found at time τ. We assume that $x(X, \tau)$ is continuous for all X and τ. In subsequent sections we shall postulate, however, that certain derivatives of $x(X, \tau)$ have jump discontinuities at the points $(Y(\tau), \tau)$. The order of the derivatives which have discontinuities at $(Y(\tau), \tau)$ determines whether the wave is a shock or an acceleration wave. The *spatial representation* $y(\tau)$, $-\infty < \tau < \infty$, of the wave is defined by

$$y(\tau) = x(Y(\tau), \tau); \tag{3.1}$$

$y(\tau)$ is the position in space of the wave at time τ.

The *wave velocity* u at time t is defined by

$$u = \frac{d}{dt} y(t) = \frac{d}{dt} x(Y(t), t); \tag{3.2}$$

u is just the rate of advance of the wave as seen by an observer at rest. We call

$$U = \frac{d}{dt} Y(t) \tag{3.3}$$

the *intrinsic velocity* of the wave. The complexity of the relation between U and u depends on how much smoothness one can assume for $x(X, t)$ as a function of X near the point $X = Y(t)$. For the moment we need only remark that, roughly speaking, U measures the velocity of propagation of the wave relative to the material.

* What we here call a *wave* is often called a "wave front". Its analogue in three-dimensional theories is sometimes called a "singular surface".

The *material trajectory* Σ of the wave is the set of ordered pairs $(Y(\tau), \tau)$ with $-\infty<\tau<\infty$. We note that the vector

$$\nu = (-1, U) \tag{3.4}$$

is normal to Σ at $(Y(t), t)$.

If $f(X, t)$, as a function of X, has only a jump discontinuity at $X = Y(t)$, then we define the *jump in* $f(X, t)$ *across* Σ *at* t by

$$[f] = \lim_{X\to Y(t)^-} f(X, t) - \lim_{X\to Y(t)^+} f(X, t). \tag{3.5}$$

Basic to the present subject is *Maxwell's Theorem* [9, § 175]: If $f(X, \tau)$ is a continuous function of X and τ jointly, while $\frac{\partial}{\partial X} f(X, \tau)$ and $\frac{\partial}{\partial \tau} f(X, \tau)$ have jump discontinuities across Σ but are continuous functions of X and τ jointly everywhere else, then the jump across Σ in the (two-dimensional) gradient of f must be parallel to ν; *i.e.*, for some scalar α

$$\left(\left[\frac{\partial f}{\partial X}\right], \left[\frac{\partial f}{\partial t}\right]\right) = (-1, U)\,\alpha; \tag{3.6}$$

or, equivalently,

$$\left[\frac{\partial f}{\partial t}\right] = -U\left[\frac{\partial f}{\partial X}\right]. \tag{3.7}$$

When f in (3.7) is the function $x(X, t)$ or one of its derivatives, the resulting equation is called a *condition of compatibility*.

In addition to satisfying kinematical conditions of compatibility a motion must obey the *law of balance of momentum*, which asserts that

$$\frac{d}{dt}\int_{X_\alpha}^{X_\beta} \dot{x}(X, t)\,\varrho_0\,dX = \int_{X_\alpha}^{X_\beta} b(X, t)\,\varrho_0\,dX + T(X_\beta, t) - T(X_\alpha, t) \tag{3.8}$$

for all times t and every pair X_α, X_β of points in $\mathscr{R}$. Here the constant ϱ_0 is the mass density in the (homogeneous) reference configuration $\mathscr{R}$, and $b(X, t)$ is the body force per unit mass. We assume that $b(X, t)$ is assigned in advance and is such that *for each* t, $b(X, t)$ *is a continuous function of* X. Of course, the stress T in (3.8) is related to motion by the constitutive equation (1.4).

4. Shock Waves

A function $f(X, s)$, defined for X in $\mathscr{R}$, $0<s<\infty$, is called *a continuous function of* X *with respect to* $\|\cdot\|$ if, for every X, $f(X, \cdot)$ belongs to $\mathscr{H}$ and

$$\lim_{\lambda\to 0} \|f(X+\lambda, \cdot) - f(X, \cdot)\| = 0.$$

We call the wave $Y(\tau)$, $-\infty<\tau<\infty$, a *shock* if the following two conditions hold.

S-1: $x(X, \tau)$ is continuous in X and τ jointly, while $F(X, \tau) = \frac{\partial}{\partial X} x(X, \tau)$ and $\dot{x}(X, \tau) = \frac{\partial}{\partial \tau} x(X, \tau)$ have jump discontinuities across Σ but are continuous in X and τ jointly everywhere else.

S-2: For each fixed t, $F^{(t)}_{(r)}(X, \cdot)$ is a continuous function of X with respect to $\|\cdot\|$.

The condition S-1 asserts that a shock is a singular surface of order 1 in the classification scheme of DUHEM and HADAMARD [9, § 187].

Since the materials we consider remember their past, we must assume that the past history of the motion is not too wild. The condition S-2 serves this purpose. When $\mathscr{H}=\mathscr{H}_h$ and $\|\cdot\|=\|\cdot\|_h$, the following conditions (1) and (2) on the motion suffice to imply S-2: (1) For each point X the set $\{\tau \mid Y(\tau)=X\}$ has Lebesgue measure zero. Roughly speaking, this means that $Y(\tau)$ does not pass through a given point too often. (2) Given any X_0 in $\mathscr{R}$ there exists a number $\lambda_0>0$ and a function $g(\cdot)$ in $\mathscr{H}_h$, possibly depending on X_0, such that $|F^{(t)}_{(r)}(X, s)|<g(s)$ for $X_0-\lambda_0\leqq X\leqq X_0+\lambda_0$ and $0<s<\infty$. In words: For each X_0, $F^{(t)}_{(r)}(X, \cdot)$ is dominated in a neighborhood of X_0 by a positive function $g(\cdot)$ in $\mathscr{H}_h$.

At each time t let us put

$$F_1=\lim_{X\to Y(t)^+} F(X,t), \qquad F_2=\lim_{X\to Y(t)^-} F(X,t), \tag{4.1a}$$

$$\dot{x}_1=\lim_{X\to Y(t)^+} \dot{x}(X,t), \qquad \dot{x}_2=\lim_{X\to Y(t)^-} \dot{x}(X,t), \tag{4.1b}$$

$$T_1=\lim_{X\to Y(t)^+} T(X,t), \qquad T_2=\lim_{X\to Y(t)^-} T(X,t). \tag{4.1c}$$

The existence of the limits in (4.1a) and (4.1b) follows immediately from the condition S-1 on the shock. The existence of the limits T_1 and T_2 follows from (1.8), the smoothness of $\mathfrak{T}(\cdot;\cdot)$, and the conditions S-1, S-2; in fact, we have

$$T_1=\mathop{\mathfrak{T}}_{s=0}^{\infty}\left(F^{(t)}_{(r)}(Y,s);F_1\right), \tag{4.2a}$$

$$T_2=\mathop{\mathfrak{T}}_{s=0}^{\infty}\left(F^{(t)}_{(r)}(Y,s);F_2\right) \tag{4.2b}$$

where $Y=Y(t)$. The jumps in F, $\dot{x}$, and T across Σ at time t are

$$[F]=F_2-F_1, \qquad [\dot{x}]=\dot{x}_2-\dot{x}_1, \qquad [T]=T_2-T_1. \tag{4.3}$$

Putting $f=x$ in (3.7) we obtain the following *compatibility condition for a shock*:*

$$[\dot{x}]=-U[F]. \tag{4.4}$$

Supposing that $Y(t)$ is between X_α and X_β, let us write (3.8) in the form

$$\begin{aligned} &T(X_\beta,t)-T(X_\alpha,t)\\ &\quad=\frac{d}{dt}\left(\int_{X_\alpha}^{Y(t)}\dot{x}(X,t)\,\varrho_0\,dX+\int_{Y(t)}^{X_\beta}\dot{x}(X,t)\,\varrho_0\,dX\right)-\int_{X_\alpha}^{X_\beta} b(X,t)\,\varrho_0\,dX. \end{aligned} \tag{4.5}$$

Clearly, if we let $X_\alpha\to Y(t)^-$ and $X_\beta\to Y(t)^+$, and use (3.3) and the assumed continuity of $b(X,t)$, we obtain from (4.5) the following expression for $[T]$:

$$[T]=-\varrho_0 U[\dot{x}]. \tag{4.6}$$

* *Cf.* TRUESDELL & TOUPIN [9, § 189], who attribute the three-dimensional form of (4.4) to HADAMARD.

Combining (4.4) and (4.6), we have*

$$U^2 = \frac{[T]}{\varrho_0 [F]}. \tag{4.7}$$

Using (4.2), (4.3), and (1.11), we can write $[T]$ in the form

$$[T] = \mathop{\mathfrak{T}}_{s=0}^{\infty}(F^{(t)}_{(r)}(Y(t), s); F_1 + [F]) - \mathop{\mathfrak{T}}_{s=0}^{\infty}(F^{(t)}_{(r)}(Y(t), s); F_1) = E_{[F]}[F] \tag{4.8}$$

where $E_{[F]}$ is the instantaneous secant modulus corresponding to the history $F^{(t)}(Y(t)^+, \cdot)$ (that is, the history just before the shock arrives) and the jump $\Lambda = [F]$. The equations (4.7) and (4.8) yield

Theorem 1. *The intrinsic velocity U of a shock satisfies*

$$U^2 = \frac{E_{[F]}}{\varrho_0} \tag{4.9}$$

where $E_{[F]}$ is the instantaneous secant modulus (1.11) *corresponding to the history just before the arrival of the shock and a jump of amount $\Lambda = [F]$.*

Suppose that $U > 0$. Then the velocity of the shock relative to the material just ahead of the shock is

$$u_1^R = u - \dot{x}_1, \tag{4.10a}$$

where $\dot{x}_1$ is given by $(4.1\text{b})_1$ and u by (3.2). Similarly, the velocity of the shock relative to the material just behind the shock is

$$u_2^R = u - \dot{x}_2. \tag{4.10b}$$

Applying a chain rule to $(3.2)_2$ and using (3.3), (4.1a) and (4.1b), we get**

$$u = F_1 U + \dot{x}_1 = F_2 U + \dot{x}_2. \tag{4.11}$$

Hence we have the following easily interpreted expressions for the intrinsic velocity:

$$U = F_1^{-1}(u - \dot{x}_1) = F_1^{-1} u_1^R, \tag{4.12a}$$

$$U = F_2^{-1}(u - \dot{x}_2) = F_2^{-1} u_2^R. \tag{4.12b}$$

5. Acceleration Waves

A function $f(X, s)$, defined for X in $\mathscr{R}$, $0 < s < \infty$, is called a *smooth function of X with respect to* $\|\cdot\|$ if, for every X, $f(X, \cdot)$ and $\frac{\partial}{\partial X} f(X, \cdot)$ belong to $\mathscr{H}$, $\lim_{\lambda \to 0} \left\| \frac{f(X+\lambda, \cdot) - f(X, \cdot)}{\lambda} - \frac{\partial}{\partial X} f(X, \cdot) \right\| = 0$ and $\frac{\partial}{\partial X} f(X, \cdot)$ is continuous in X with respect to $\|\cdot\|$.

---

* For the special form taken by (4.7) for one-dimensional shocks in perfect fluids see eq. (54.9) of [*10*], which equation has been known a long time and is often associated with Hugoniot. Several formulae equivalent to our (4.7) are listed by Truesdell & Toupin [*9*]; see in particular their eqs. (205.3), (205.4), (205.6), (205.7), and the historical comments on their p. 545.

** It will no doubt be noticed that (4.11) furnishes another derivation of the compatibility condition (4.4).

We call $Y(\tau)$, $-\infty<\tau<\infty$, an *acceleration wave* if the following two conditions hold.

A-1: $x(X,\tau)$, $F(X,\tau)$, and $\dot{x}(X,\tau)$ are continuous in X and τ jointly, while $\frac{\partial}{\partial X}F(X,\tau)$, $\dot{F}(X,\tau)$, $\frac{\partial}{\partial X}\dot{x}(X,\tau)$, and $\ddot{x}(X,\tau)$ have jump discontinuities across Σ but are continuous in X and τ jointly everywhere else.

A-2: For each fixed t, $F^{(t)}_{(r)}(X,\cdot)$ is a smooth function of X with respect to $\|\cdot\|$.

The condition A-1 asserts that an acceleration wave is a Duhem-Hadamard singular surface of order 2, [9, § 187].

The condition A-2 limits the wildness of the past history of a point. When $\mathscr{H}=\mathscr{H}_h$ and $\|\cdot\|=\|\cdot\|_h$, the following conditions imply A-2: (1) For each X, $\{\tau\,|\,Y(\tau)=X\}$ has Lebesgue measure zero. (2) Given X_o in $\mathscr{R}$ there exists a number $\lambda_o>0$ and a function $g(\cdot)$ in $\mathscr{H}_h$ such that

$$|F^{(t)}_{(r)}(X,s)|<g(s) \quad \text{and} \quad \left|\frac{\partial}{\partial X}F^{(t)}_{(r)}(X,s)\right|<g(s) \quad \text{for} \quad X_o-\lambda_o\leqq X\leqq X_o+\lambda_o$$

and $0<s<\infty$.

Putting $f=\dot{x}$ and $f=F$ in (3.7), we obtain

$$[\ddot{x}]=-U\left[\frac{\partial\dot{x}}{\partial X}\right], \tag{5.1 a}$$

$$[\dot{F}]=-U\left[\frac{\partial F}{\partial X}\right], \tag{5.1 b}$$

and, since $\left[\frac{\partial\dot{x}}{\partial X}\right]=[\dot{F}]$, we have the following well known *compatibility condition for an acceleration wave*:*

$$[\ddot{x}]=U^2\left[\frac{\partial F}{\partial X}\right]. \tag{5.2}$$

Now, it follows from our basic constitutive equation (1.8), our smoothness hypotheses on the functional $\mathfrak{T}$, and our definition of an acceleration wave, that $\frac{\partial}{\partial X}T(X,t)$ exists for each t and is continuous in X on $\mathscr{R}$ except for a jump discontinuity at $Y(t)$. Hence, (3.8) here becomes equivalent to the assertion that whenever $X\neq Y(t)$ the differential equation,

$$\frac{\partial}{\partial X}T+\varrho_o b=\varrho_o\ddot{x}, \tag{5.3}$$

holds and that the jumps in $\frac{\partial}{\partial X}T$ and $\ddot{x}$ across Σ at time t are related by

$$\left[\frac{\partial T}{\partial X}\right]=\varrho_o[\ddot{x}]. \tag{5.4}$$

Furthermore, whenever $X\neq Y(t)$, (2.2), (2.7), (A-1), and (A-2) imply the chain rule:

$$\begin{aligned}\frac{\partial}{\partial X}T(X,t)&=\delta\,\underset{s=0}{\overset{\infty}{\mathfrak{T}}}\Big(F^{(t)}_{(r)}(X,s);F(X,t)\,\Big|\,\frac{\partial}{\partial X}F^{(t)}_{(r)}(X,s)\Big)+\\&\quad+D\,\underset{s=0}{\overset{\infty}{\mathfrak{T}}}\big(F^{(t)}_{(r)}(X,s);F(X,t)\big)\frac{\partial}{\partial X}F(X,t).\end{aligned} \tag{5.5}$$

---

* For a thorough discussion of the situation in three-dimensions see TRUESDELL & TOUPIN [9, § 190]; they attribute the basic results to HUGONIOT and HADAMARD.

But $F(X,t)$ is a continuous function of X, while $\frac{\partial}{\partial X} F^{(t)}_{(r)}(X,\cdot)$ and $F^{(t)}_{(r)}(X,\cdot)$ are continuous functions of X with respect to $\|\cdot\|$. Indeed, the continuity of the first two quantities is assumed in A-1 and A-2; the continuity of the third follows from that of the second. Thus, by the smoothness of $\mathfrak{T}(\cdot;\cdot)$ we have that $\delta \mathop{\mathfrak{T}}\limits_{s=0}^{\infty}\left(F^{(t)}_{(r)}(X,s); F(X,t) \,\middle|\, \frac{\partial}{\partial X} F^{(t)}_{(r)}(X,s)\right)$ and the tangent modulus $E = D \mathop{\mathfrak{T}}\limits_{s=0}^{\infty}\left(F^{(t)}_{(r)}(X,s); F(X,t)\right)$ are also continuous in X, for all X. Consequently, (5.5) implies that

$$\left[\frac{\partial T}{\partial X}\right] = E\left[\frac{\partial F}{\partial X}\right]. \tag{5.6}$$

Combining this equation with (5.4) and (5.2), we find that

$$(\varrho_o U^2 - E)\left[\frac{\partial F}{\partial X}\right] = 0.$$

We have just proven

Theorem 2. *The intrinsic velocity U of an acceleration wave satisfies*

$$U^2 = \frac{E}{\varrho_o} \tag{5.7}$$

where E is the instantaneous tangent modulus (1.10) *corresponding to the history* $F^{(t)}(Y(t),\cdot)$.

The relative velocity of an acceleration wave is defined by*

$$u^R = u - \dot{x}(Y(t), t) \tag{5.8}$$

where u is the wave velocity defined in (3.2). Applying a familiar chain rule to $(3.2)_2$ and using (3.3), we get

$$u = F U + \dot{x} \tag{5.9}$$

where $F = F(Y(t), t)$ and $\dot{x} = \dot{x}(Y(t), t)$. Hence,

$$U = F^{-1} u^R \tag{5.10}$$

and (5.7) may be written

$$(u^R)^2 = F^2 \frac{E}{\varrho_o}. \tag{5.11}$$

It follows from (5.7) and (2.16a) that in the theory of linear viscoelasticity

$$U^2 = G(0)/\varrho_o. \tag{5.12}$$

When an acceleration wave enters a general simple material which has always been at rest in its reference configuration, $F^{(t)}(Y(t),\cdot)$ is such that $F(t)=1$, $F^{(t)}_{(r)}(\cdot) = 1^{\dagger}_{(r)}(\cdot)$, and hence E in (5.7) is given by

$$E = D \mathop{\mathfrak{T}}\limits_{s=0}^{\infty}\left(1^{\dagger}_{(r)}(s); 1\right). \tag{5.13}$$

Suppose now that this simple material obeys the principle of fading memory. Then, by $(2.24)_2$, E in (5.13) is just $G(0)$, *i.e.*, the initial value of $G(\cdot)$ for the linearly viscoelastic material which approximates the general simple material in small

---

* *Cf.* (4.10).

strain as explained in the remark at the end of Section 2. Furthermore, here we not only have $F=1$ but also $\dot{x}=0$ in (5.9), and hence $u=U$. This proves

Theorem 3. *Consider an acceleration wave entering a general simple material obeying the principle of fading memory and suppose that the material in front of the wave has always been at rest in the configuration $\mathcal{R}$. Then regardless of the amplitude of the wave its velocity obeys the formula* (5.12) *of linear viscoelasticity, with* $G(0)$ *the initial value of the relaxation function of that linearly viscoelastic material which approximates the general simple material in small deformations from $\mathcal{R}$. Furthermore, the intrinsic velocity U in the formula* (5.12) *is then equal to the ordinary wave velocity u.*

6. Examples of One-Dimensional Motions

Consider a three-dimensional homogeneous isotropic body $\mathcal{B}$ and a fixed Cartesian coordinate system. Let x^i, $i=1, 2, 3$, be the coordinates at time τ, $-\infty<\tau<\infty$, of that material point of $\mathcal{B}$ which has the coordinates X^j when $\mathcal{B}$ is in a fixed homogeneous and undistorted reference configuration $\mathcal{R}$. We classify motions of $\mathcal{B}$ according to the dependence of x^i on X^j and τ. The deformation gradient is the tensor with Cartesian components

$$F^{ij}(X^k, \tau) = \frac{\partial}{\partial X^j}\, x^i(X^k, \tau). \tag{6.1}$$

In a compressible isotropic material the stress tensor T^{ij} is given by

$$T^{ij} = T^{ij}(X^k, t) = \mathop{\mathfrak{T}^{ij}}_{s=0}^{\infty}\left(F^{kl}(X^m, t-s)\right); \tag{6.2}$$

where $\mathfrak{T}^{ij}$ is an isotropic tensor-valued functional.* If the isotropic material is incompressible then F^{ij} obeys the constraint

$$|\det F^{ij}| = 1$$

and T^{ij} is determined by the history of F^{ij} only to within a hydrostatic pressure $-p\,\delta^{ij}$:

$$T^{ij} = -p(X^k, t)\,\delta^{ij} + \mathop{\mathfrak{T}_{\mathsf{E}}^{ij}}_{s=0}^{\infty}\left(F^{lm}(X^k, t-s)\right). \tag{6.3}$$

The value T_{E}^{ij} of the isotropic functional $\mathfrak{T}_{\mathsf{E}}^{ij}$ is called the extra-stress. Without loss of generality it may be normalized so that

$$T_{\mathsf{E}}^{ii} = 0, \qquad \text{i.e.,} \qquad p = -\tfrac{1}{3} T^{ii}. \tag{6.4}$$

The *law of balance of momentum* asserts that in every part** $\mathcal{P}$ of $\mathcal{B}$ the following relation must hold at each time t:

$$\frac{d}{dt}\int_{\mathcal{P}} \dot{x}^i \varrho\, dv = \int_{\mathcal{P}} b^i \varrho\, dv + \int_{\partial\mathcal{P}} T^{ij}\, da_j, \tag{6.5}$$

where dv is the element of volume, ϱ and b^i the mass density and body force density at time t, and da_j the element of area of the surface $\partial\mathcal{P}$.

---

* We do not distinguish between a tensor functional and its component functionals. For the definition of an *isotropic functional* see, for example, eq. (2.2) of [*11*].

** By a part $\mathcal{P}$ of $\mathcal{B}$ we mean a simply connected bounded set which is the closure of an open subset of $\mathcal{R}$ and has a piecewise smooth boundary $\partial\mathcal{P}$.

A. Longitudinal Stretch of a Cylinder

Let the compressible body $\mathcal{B}$ be such that when $\mathcal{B}$ is in its reference configuration it has the form of a cylinder whose axis is along the x^1-axis of our Cartesian coordinate system. We say that the motion of $\mathcal{B}$ is *purely longitudinal* if

$$\begin{aligned} x^1 &= \omega(X, \tau), \\ x^2 &= X^2, \\ x^3 &= X^3 \end{aligned} \tag{6.6}$$

and

$$\omega' = \frac{\partial \omega}{\partial X^1}(X^1, \tau) \neq 0. \tag{6.7}$$

Notice that in such a motion

$$\dot{x}^1 = \frac{\partial}{\partial \tau}\omega(X^1, \tau), \qquad \dot{x}^2 = 0, \qquad \dot{x}^3 = 0, \tag{6.8}$$

$$[F^{ij}] = \begin{bmatrix} \omega'(X^2, \tau) & 0 & 0 \\ 0 & 1 & 0 \\ 0 & 0 & 1 \end{bmatrix} \tag{6.9}$$

and the body remains a cylinder at all times. It is shown in [*12*] that in an isotropic body the stress tensor corresponding to a motion of the type (6.6) has the form

$$[T^{ij}] = \begin{bmatrix} T^{11} & 0 & 0 \\ 0 & T^{22} & 0 \\ 0 & 0 & T^{33} \end{bmatrix} \tag{6.10}$$

where

$$T^{11} = T^{11}(X^1, t) = \underset{s=0}{\overset{\infty}{\mathfrak{g}}}\big(\omega'(X^1, t-s)\big), \tag{6.11}$$

$$T^{22} = T^{33} = T^{33}(X^1, t) = \underset{s=0}{\overset{\infty}{\mathfrak{h}}}\big(\omega'(X^1, t-s)\big). \tag{6.12}$$

Here $\mathfrak{g}$ and $\mathfrak{h}$ are scalar-valued functionals with argument $\omega'(x^1, t-\cdot)$. These functionals $\mathfrak{g}$ and $\mathfrak{h}$ are determined by the functional $\mathfrak{T}^{ij}$ of (6.2), *i.e.*, by the material under consideration, and hence are called *material functionals.*

We assume now that the body force has the form

$$b^1 = b = b(X^1, t), \qquad b^2 = b^3 = 0. \tag{6.13}$$

Using the fact that here T^{ij} has the form (6.10) with components that are functions of only X^1 and t (and hence of only x^1 and t) one can easily show that for each part $\mathcal{P}$ of $\mathcal{B}$

$$\int_{\partial\mathcal{P}} T^{2j}\,da_j = \int_{\partial\mathcal{P}} T^{22}\,da_2 = 0, \tag{6.14}$$

$$\int_{\partial\mathcal{P}} T^{3j}\,da_j = \int_{\partial\mathcal{P}} T^{33}\,da_3 = 0. \tag{6.15}$$

Hence, by (6.8) and (6.13), the two equations obtained from (6.5) by putting $i=2, 3$ now both reduce to $0=0$. The reader will have no trouble convincing himself that the remaining equation is satisfied for all $\mathcal{P}$ if and only if for all numbers α and β we have

$$\frac{d}{dt}\int_{\alpha}^{\beta} \dot{x}^1 \varrho\, dx^1 = \int_{\alpha}^{\beta} b\, \varrho\, dx^1 + T^{11}|_{x^1=\beta} - T^{11}|_{x^1=\alpha}. \tag{6.16}$$

Let us now put

$$T = T^{11}, \qquad X = X^1, \qquad x = x^1. \tag{6.17}$$

$$\varrho_0 = \varrho\, \omega' = \varrho \det[F_{ij}], \tag{6.18}$$

$$F(X, \tau) = \omega'(X^1, \tau) = \frac{\partial}{\partial X} x(X, \tau). \tag{6.19}$$

Here T is just the tension in the cylinder, X the distance along the cylinder, and ϱ_0 the mass density in the reference configuration. It follows from (6.16)—(6.19) that for an isotropic cylinder engaged in a longitudinal motion the law of balance of momentum (6.5) holds if and only if the equation

$$\frac{d}{dt}\int_{X_\alpha}^{X_\beta} \dot{x}(X, t)\, \varrho_0\, dX = \int_{X_\alpha}^{X_\beta} b(X, t)\, \varrho_0\, dX + T(X_\beta, t) - T(X_\alpha, t) \tag{6.20}$$

holds at all times t and for all numbers X_α and X_β.

Equation (6.20) is identical to the law of balance of momentum (3.8) assumed in our one-dimensional theory. It is now clear that the one-dimensional theory describes longitudinal waves propagating along a cylinder provided we make the identifications (6.17)—(6.19) and equate the functional $\mathfrak{T}$ of (1.4) with the functional $\mathfrak{g}$ of (6.11).

Although it is satisfying to have three-dimensional motions which are precisely described in our one-dimensional framework, we must emphasize the practical difficulties involved in achieving purely longitudinal motions in the laboratory. In a motion of the type (6.6) the surface of the cylindrical body must be free to move in the x^1 direction but must be constrained in such a way that transverse motions are impossible. The contact forces required to oppose transverse motions are determined by the functional $\mathfrak{h}$ in (6.12). These forces, which are normal to the cylindrical surface, not only vary with the material and the motion, but also depend on time and the position X^1 along the cylinder in a given motion. For a circular cylinder, the normal thrust per unit area that must be applied to the bounding cylindrical surface is equal to $\mathop{\mathfrak{g}}\limits_{s=0}^{\infty}(\omega'(X^1, t-s))$. Even if the material is a solid, it will not do to let the cylindrical surface be completely free or subject to only a fixed uniform hydrostatic pressure. It may be that the errors made in applying our one-dimensional analysis to a wave traveling down a finite cylinder with completely free surface are unimportant when the diameter is very large compared to the product of the wave velocity with the time since impact, but a rigorous analysis to this effect is lacking. It may also be possible for an experimenter to design a lubricated sleeve which when placed over the cylindrical specimen leaves longitudinal motions unhindered but prohibits transverse motions, but this, too, is mere conjecture.

We remark, but omit the proof, that all of the results we have given here for isotropic materials in longitudinal motion hold also for transversely isotropic materials in such motion provided only that the direction x^1 of the motion is made to correspond with the axis of symmetry of the material.

A perfect fluid (with no thermodynamic variables) corresponds to the choice

$$T^{ij} = -p(\det[F^{kl}])\, \delta^{ij} \tag{6.21}$$

for the functional $\mathfrak{T}^{ij}$ in (6.2). For that special isotropic material our present theory of longitudinal motion reduces to the familiar theory of one-dimensional flow down a pipe. The boundary conditions required by (6.8), *i.e.*, the conditions that the velocity have no component normal to the surface of the pipe but be unrestricted for tangential motions, are regarded as natural in the theory of perfect fluids.

B. Shearing Motions

We here suppose that the isotropic body $\mathscr{B}$ is incompressible. We say that the motion of $\mathscr{B}$ is a *pure shear* if in some Cartesian coordinate system

$$\begin{aligned} x^1 &= X^1, \\ x^2 &= X^2 + \lambda(X^1, \tau), \\ x^3 &= X^3. \end{aligned} \tag{6.22}$$

For such a motion

$$\dot{x}^1 = 0, \qquad \dot{x}^2 = \frac{\partial}{\partial \tau}\lambda(X^1, \tau), \qquad \dot{x}^3 = 0 \tag{6.23}$$

and

$$[F^{ij}] = \begin{bmatrix} 1 & 0 & 0 \\ \lambda'(X^1, \tau) & 1 & 0 \\ 0 & 0 & 1 \end{bmatrix}, \qquad \lambda' = \frac{\partial}{\partial X^1}\lambda. \tag{6.24}$$

In an incompressible isotropic material the extra-stress corresponding to a motion of the type (6.22) has the form*

$$[T_E^{ij}] = \begin{bmatrix} T_E^{11} & T_E^{21} & 0 \\ T_E^{12} & T_E^{22} & 0 \\ 0 & 0 & T_E^{33} \end{bmatrix}, \tag{6.25}$$

with its nonzero components determined by three functionals $\mathfrak{t}$, $\mathfrak{s}_1$ and $\mathfrak{s}_2$ as follows:

$$\begin{aligned} T_E^{12} = T_E^{21} &= \mathop{\mathfrak{t}}_{s=0}^{\infty}(\lambda'(X^1, t-s)) \\ T_E^{11} - T_E^{22} &= \mathop{\mathfrak{s}_1}_{s=0}^{\infty}(\lambda'(X^1, t-s)) \\ T_E^{22} - T_E^{33} &= \mathop{\mathfrak{s}_2}_{s=0}^{\infty}(\lambda'(X^1, t-s)) \end{aligned} \tag{6.26}$$

$$T_E^{11} + T_E^{22} + T_E^{33} = 0. \tag{6.27}$$

The material functionals $\mathfrak{t}, \mathfrak{s}_1, \mathfrak{s}_2$ are determined by the functional $\mathfrak{T}_E^{ij}$ of (6.3) and vary from substance to substance. They obey the identities

$$\mathop{\mathfrak{t}}_{s=0}^{\infty}(-\lambda'(t-s)) = -\mathop{\mathfrak{t}}_{s=0}^{\infty}(\lambda'(t-s)), \qquad \mathop{\mathfrak{s}_i}_{s=0}^{\infty}(-\lambda'(t-s)) = \mathop{\mathfrak{s}_i}_{s=0}^{\infty}(\lambda'(t-s)). \tag{6.28}$$

Equation (6.27) is merely our normalization convention $(6.4)_1$. Notice that it follows from (6.25)—(6.27) that T_E^{ij} is a function of X^1 and t alone.

* For incompressible fluids this was shown in [*11*]; in [*12*] it is observed that the result holds also for incompressible isotropic simple materials in general.

Let us now assume that the continuous body force field b^i has a single-valued potential ψ; *i.e.*,

$$b^i = -\frac{\partial \psi}{\partial x^i}, \tag{6.29}$$

and put

$$\varphi = p + \varrho\,\psi. \tag{6.30}$$

We consider the special case in which p and ψ are such that

$$\frac{\partial \varphi}{\partial x^2} = 0. \tag{6.31}$$

When $b^2 = 0$, the assumption of (6.31) is equivalent to assuming that the pressure field p has no gradient in the flow direction.

Using (6.25), (6.26) and (6.30), one can show that the law of balance of momentum (6.5) is equivalent to the following two equations:

$$\frac{d}{dt}\int_\alpha^\beta \dot{x}^2 \varrho\, dx^1 = T^{21}\big|_{x^1=\beta} - T^{21}\big|_{x^1=\alpha} \tag{6.32}$$

$$\varphi = T^{11} - h(t). \tag{6.33}$$

Equation (6.32) must hold for all choices of the numbers α and β such that $x^1 = \alpha$ and $x^1 = \beta$ correspond to points in the body. In (6.32) h denotes an arbitrary function of time. Once a motion of the type (6.22) is found satisfying (6.32), T^{11} and all the other components of the extra-stress are determined, and (6.33) may be used to calculate the spatial dependence of φ. Hence, the crucial problem here is that of solving (6.32) for the function λ, when T^{21} is given by $(6.26)_1$. If, however, we put

$$\left.\begin{aligned}
&T = T^{21}, \qquad X = X^1 = x^1, \qquad x = x^2 - X^2 + X^1 = \lambda(X^1, \tau) + X^1 = x(X, t),\\
&F(X, \tau) = \frac{\partial}{\partial X} x(X, \tau) = \lambda'(X^1, \tau) + 1,\\
&\varrho_0 = \varrho,\\
&\mathfrak{T}_{s=0}^{\infty}\left(F^{(t)}(s)\right) = \mathfrak{t}_{s=0}^{\infty}\left(\lambda'(t-s)\right); \text{ i.e., } \mathfrak{T}_{s=0}^{\infty}\left(F^{(t)}(s)\right) = \mathfrak{t}_{s=0}^{\infty}\left(F^{(t)}(s) - 1^\dagger(s)\right),
\end{aligned}\right\} \tag{6.34}$$

we see that solving (6.32) subject to $(6.26)_2$ is the same as solving (3.8) with $b = 0$ and T given by (1.4). Hence our one-dimensional theory applies to the shearing of an isotropic incompressible material whenever $\frac{\partial}{\partial x^2}\varphi = 0$.

Note that the waves described by our present interpretation of the one-dimensional theory are not longitudinal but rather transverse. Hence, in this interpretation those waves which we previously called *shocks* are really *vortex sheets.** Furthermore, in place of the equations (4.10)—(4.12) and (5.8)—(5.10) we here have $U = u$ for both shock and acceleration waves.

To achieve shearing motion in the laboratory one might use a sample shaped as a parallelepiped with one pair of faces $\mathscr{S}_1$ and $\mathscr{S}_2$ perpendicular to the x^1-plane. The experimenter would try to generate a wave by suddenly moving one of these surfaces in the x^2-direction. Unfortunately, to maintain a shearing motion the remaining

* [9, §§ 185, 189].

surfaces could not, in general, be left free to the ambient atmosphere but would have to receive tractions varying in a complicated way with time and position. It may be possible to make negligible the error involved in leaving free the side surfaces by choosing the area of $\mathscr{S}_1$ and $\mathscr{S}_2$ to be relatively large, but we have no mathematical argument justifying this intuitive supposition.

Acknowledgements. We thank Professor ELLIS H. DILL for valuable discussions about the material of Section 6. This research was supported in part by the following agencies of the United States Government: The National Science Foundation, the Office of Naval Research, and the Advanced Research Projects Agency of the Department of Defense.

References

[1] NOLL, W., Arch. Rational Mech. Anal. **2**, 197—226 (1958).
[2] COLEMAN, B. D., & W. NOLL, Arch. Rational Mech. Anal. **6**, 355—370 (1960).
[3] COLEMAN, B. D., & W. NOLL, Rev. Modern Phys. **33**, 239—249 (1961).
[4] COLEMAN, B. D., & W. NOLL, Internat. Sympos. Second-Order Effects, Haifa, 1962: Macmillan 1964, pp. 530—552.
[5] FISHER, G. M. C., & M. E. GURTIN, Quart. Appl. Math. (1965).
[6] HERRERA R., I., & M. E. GURTIN, Quart. Appl. Math. **22**, 360—364 (1965).
[7] COLEMAN, B. D., Arch. Rational Mech. Anal. **17**, 1—46 (1964).
[8] COLEMAN, B. D., Arch. Rational Mech. Anal. **17**, 230—254 (1964).
[9] TRUESDELL, C., & R. A. TOUPIN, The Classical Field Theories. In the Encyclopedia of Physics, Vol. III/1, edited by S. FLÜGGE. Berlin-Göttingen-Heidelberg: Springer 1960.
[10] SERRIN, J., Mathematical Principles of Classical Fluid Mechanics. In the Encyclopedia of Physics, Vol. VIII/1, edited by S. FLÜGGE. Berlin-Göttingen-Heidelberg: Springer 1959.
[11] COLEMAN, B. D., & W. NOLL, Ann. N.Y. Acad. Sci. **89**, 672—714 (1961).
[12] COLEMAN, B. D., Arch. Rational Mech. Anal., forthcoming.

Mellon Institute, Pittsburgh
Brown University, Providence
and
University of Mexico, Mexico City

(Received October 22, 1964)

Waves in Materials with Memory

II. On the Growth and Decay of One-Dimensional Acceleration Waves

BERNARD D. COLEMAN & MORTON E. GURTIN

Contents

Preface

Paper I of this series laid the foundations of a theory of one-dimensional shock and acceleration waves in general nonlinear materials with memory and gave formulae for the velocity of such waves [*1*].★ There the reader will find a detailed explanation of the terminology to be used here and examples of circumstances in which the behavior of three-dimensional bodies is described by the one-dimensional theory.

In the present article we derive general and explicit expressions for the change in amplitude of one-dimensional acceleration waves propagating through arbitrary homogeneous simple materials obeying the principle of fading memory.

1. The Fading Memory Assumption

Since our present results rest on smoothness hypotheses stronger than those used in I, we here briefly recapitulate the definition of a one-dimensional homogeneous simple material with fading memory, strengthening it where necessary.

The stress T at time t in a *simple material* is determined by the history F^t of the deformation gradient:

$$T(t) = \mathfrak{T}(F^t). \tag{1.1}$$

★ Henceforth we shall refer to reference [*1*] simply as "I".

Here F^t is the function defined by*

$$F^t(s) = F(t-s) = \frac{\partial}{\partial X} x(X, t-s), \quad 0 \leqq s < \infty, \tag{1.2}$$

with $x(X, t-s)$ the location at time $t-s$ of the material point which has the location X in a homogeneous reference configuration $\mathscr{R}$. The assumptions that the material is *homogeneous* and that a homogeneous configuration has been taken as reference are equivalent to the following two assertions: (1) Although F^t and $T(t)$ depend on X, the functional $\mathfrak{T}$ in (1.1) is independent of X. (2) The mass density ϱ_R in the reference configuration is independent of X.

A knowledge of F^t, whose domain of definition is $[0, \infty)$, is equivalent to knowledge of the present value $F = F^t(0)$ and the past history F_r^t, where F_r^t is the restriction of F^t to $(0, \infty)$. Hence, we can write (1.1) in the form

$$T(t) = \mathfrak{T}(F_r^t; F). \tag{1.3}$$

At the risk of repetition, we pause here to emphasize the differences in the mathematical character of F, F^t, and F_r^t: the *present value* $F = F(t) = F^t(0)$ is a *number*; the *history* F^t is a *function* whose domain is $[0, \infty)$ and whose value at $s \geqq 0$ is $F^t(s) = F(t-s)$; the *past history* F_r^t is a *function* whose domain is $(0, \infty)$ and whose value for each $s > 0$ is $F_r^t(s) = F^t(s)$.

Let h be a fixed *influence function*, i.e., a positive, monotone-decreasing, continuous function on $[0, \infty)$ decaying to zero fast enough to be square integrable, and put

$$\|F_r^t\|_h = \sqrt{\int_0^\infty |F_r^t(s)|^2 h(s)^2 \, ds}, \tag{1.4}$$

$$\|F^t\|_h^\oplus = \|F_r^t\|_h + |F|, \tag{1.5}$$

where $|\cdot|$ denotes the absolute value, and, again, $F = F^t(0)$. Let $\mathscr{H}_h$ be the set of real functions g on $(0, \infty)$ with $\|g\|_h$ finite and $\mathscr{H}_h^\oplus$ the set of real functions f on $[0, \infty)$ with $\|f\|_h^\oplus$ finite. Our *assumption of fading memory* asserts that there must exist an influence function h such that $\mathfrak{T}(\cdot)$ in (1.1) has for its domain $\mathscr{D}$ an open subset of $\mathscr{H}_h^\oplus$ and is a smooth function, throughout $\mathscr{D}$, with respect to the norm $\|\cdot\|_h^\oplus$ of (1.5). More precisely, we assume that for each fixed history F^t in $\mathscr{D}$ there exist n^{th}-order Fréchet derivatives $\delta^{\oplus n}\mathfrak{T}(F^t|\cdot, \ldots, \cdot)$ of $\mathfrak{T}(\cdot)$ for $n = 1, 2, \ldots, N$. These derivatives are, by definition, bounded, symmetric, n-linear forms over $\mathscr{H}_h^\oplus$ and have the property that

$$\mathfrak{T}(F^t + f) = \mathfrak{T}(F^t) + \sum_{n=1}^{N} \frac{1}{n!} \delta^{\oplus n} \mathfrak{T}(F^t | f, f, \ldots, f) + o\left((\|f\|_h^\oplus)^N\right). \tag{1.6}$$

* It will be noticed that the notation we use here resembles that introduced by NOLL [2] and is very condensed compared to that used in I. Our present F^t and $\mathfrak{T}(F^t)$ were written, respectively, $F^{(t)}(\cdot)$ and $\underset{s=0}{\overset{\infty}{\mathfrak{T}}}(F^{(t)}(s))$ in I. Although the notation of I has the advantage of emphasizing that F^t is a function and that $T = \mathfrak{T}(F^t)$ may depend on all the values $F^t(s)$ of F^t, if used here, our previous notation would soon yield formulae of such great length and containing so many parentheses as to be nearly unintelligible. Furthermore, as tensorial indices do not occur on F in this article, we do not need to enclose in parentheses the t in F^t.

Furthermore, we assume that each of the functionals $\delta^{\oplus n}\mathfrak{T}(\cdot|\cdot,\dots,\cdot)$ is jointly continuous with respect to the product topology in all of its $n+1$ arguments. For the results obtained in I it sufficed to assume that $N=1$. Here we assume that $N=2$; this strengthened assumption of smoothness insures the existence and continuity of the following "partial derivatives" of $\mathfrak{T}$:

$$D\mathfrak{T}(F^t) \overset{\text{def}}{=} \frac{\partial}{\partial F}\mathfrak{T}(F_r^t; F), \tag{1.7a}$$

$$D^2\mathfrak{T}(F^t) \overset{\text{def}}{=} \frac{\partial^2}{\partial F^2}\mathfrak{T}(F_r^t; F), \tag{1.7b}$$

$$\delta\mathfrak{T}(F^t|g) \overset{\text{def}}{=} \frac{\partial}{\partial \alpha}\mathfrak{T}(F_r^t + \alpha g; F)\big|_{\alpha=0}, \tag{1.7c}$$

$$\delta^2\mathfrak{T}(F^t|g, l) \overset{\text{def}}{=} \frac{\partial^2}{\partial\alpha\,\partial\beta}\mathfrak{T}(F_r^t + \alpha g + \beta l; F)\Big|_{\substack{\alpha=0\\ \beta=0}} = \delta^2\mathfrak{T}(F^t|l, g), \tag{1.7d}$$

$$\begin{aligned} \delta D\mathfrak{T}(F^t|g) &\overset{\text{def}}{=} \frac{\partial^2}{\partial\alpha\,\partial F}\mathfrak{T}(F_r^t + \alpha g; F)\big|_{\alpha=0} \\ &= \frac{\partial^2}{\partial F\,\partial\alpha}\mathfrak{T}(F_r^t + \alpha g; F)\big|_{\alpha=0} \overset{\text{def}}{=} D\delta\mathfrak{T}(F^t|g). \end{aligned} \tag{1.7e}$$

These equations must hold for all histories F^t in $\mathscr{D}$ and functions g, l in $\mathscr{H}_h$.

In I we called the number

$$E_t = D\mathfrak{T}(F^t) \tag{1.8}$$

the instantaneous tangent modulus corresponding to the history F^t; it equals the initial slope of the stress-strain curve for instantaneous response to strain impulses superimposed on F^t at time t.* The number

$$\tilde{E}_t = D^2\mathfrak{T}(F^t), \tag{1.9}$$

which we call the *instantaneous second-order modulus* for F^t, measures the initial curvature of the stress-strain curve for instantaneous response.

The "partial derivatives" defined in (1.7) are related to the "total derivatives" in (1.6) by the following identities which hold for all functions f and p in $\mathscr{H}_h^{\oplus}$:

$$\delta^{\oplus}\mathfrak{T}(F^t|f) = \delta\mathfrak{T}(F^t|f_r) + D\mathfrak{T}(F^t)\,f(0), \tag{1.10a}$$

$$\begin{aligned} \delta^{\oplus 2}\mathfrak{T}(F^t|f, p) = \delta^2\mathfrak{T}(F^t|f_r, p_r) + D\delta\mathfrak{T}(F^t|f_r)\,p(0) + \\ + \delta D\mathfrak{T}(F^t|p_r)\,f(0) + D^2\mathfrak{T}(F^t)\,f(0)\,p(0). \end{aligned} \tag{1.10b}$$

Here f_r and p_r are the restrictions of f and p to $(0, \infty)$.

Clearly, by (1.7) and the assumption asserted after (1.6), the functionals $\delta^n\mathfrak{T}(F^t|\cdot,\dots,\cdot)$ $(n=1, 2)$ are bounded, symmetric, n-linear forms over the Hilbert space $\mathscr{H}_h$. In particular, $\delta\mathfrak{T}(F^t|g)$, as a function of g, is both linear and continuous with respect to $\|\cdot\|_h$, and, therefore, can be represented as an inner-product with a function K_t in $\mathscr{H}_h$:

$$\delta\mathfrak{T}(F^t|g) = \int_0^\infty K_t(s)\,g(s)\,h(s)^2\,ds. \tag{1.11}$$

Of course, K_t depends on F^t; putting

$$G_t'(s) = K_t(s)\,h(s)^2, \tag{1.12}$$

* For a more complete discussion see [3].

we can make this dependence explicit by writing

$$G'_t(s) = \mathfrak{G}'(F^t; s). \tag{1.13}$$

We assume that $\mathfrak{G}'(\cdot, s)$ is, for each fixed s, a continuous functional over $\mathscr{D}$ and that, for each fixed history F^t in $\mathscr{D}$, $G'_t(s) = \mathfrak{G}'(F^t, s)$ is a differentiable function of s for $0 \leqq s < \infty$. Putting $G''_t(s) = \frac{d}{ds} G'_t(s)$, we further assume that the mapping $F^t(\cdot) \to G''_t(\cdot)\, h(\cdot)^{-2}$ carries $\mathscr{D}$ into $\mathscr{H}_h$ and is continuous. To emphasize the dependence of G''_t on F^t, as in (1.13) we write

$$G''_t(s) = \mathfrak{G}''(F^t; s). \tag{1.14}$$

Let us consider the unique solution G_t of

$$\frac{d}{ds} G_t(s) = G'_t(s), \tag{1.15a}$$

$$G_t(0) = E_t = D\mathfrak{T}(F^t). \tag{1.15b}$$

Of course, this integral G_t of G'_t also depends on F^t:

$$G_t(s) = \mathfrak{G}(F^t; s). \tag{1.16}$$

G_t may be called the *stress-relaxation function corresponding to the history* F^t. It follows from the definition of G_t and (1.10a) that for each F^t in $\mathscr{D}$ and each f in $\mathscr{H}_h^{\oplus}$,

$$\delta^{\oplus}\mathfrak{T}(F^t | f) = G_t(0)\, f(0) + \int_0^\infty G'_t(s)\, f(s)\, ds. \tag{1.17}$$

Hence, G_t supplies a generalization, useful for simple materials with arbitrary history, of the stress-relaxation function G of linear viscoelasticity. This is borne out by the following extension of the remark made at the end of Section 2 of I.

Remark 1.1. Let $F(\tau)$ and $\tilde{F}(\tau)$, with $-\infty < \tau < \infty$, be such that for each t the corresponding histories up to time t, F^t and $\tilde{F}^t$, are in the domain $\mathscr{D}$ of $\mathfrak{T}$. Suppose the *relative strain* defined by

$$\gamma(\tau) = \tilde{F}(\tau) - F(\tau) \tag{1.18}$$

is always small. Then, if we put

$$T(t) = \mathfrak{T}(F^t), \qquad \tilde{T}(t) = \mathfrak{T}(\tilde{F}^t), \tag{1.19}$$

and

$$\varepsilon = \sup_{-\infty < \tau < \infty} |\gamma(\tau)|, \tag{1.20}$$

we have, for all t,

$$\tilde{T}(t) = T(t) + G_t(0)\, \gamma(t) + \int_0^\infty G'_t(s)\, \gamma(t-s)\, ds + o(\varepsilon) \tag{1.21}$$

where G_t is the stress-relaxation function corresponding to F^t.

This remark is an immediate consequence of the equation (1.17), the equation (1.6) with $N=1$, and the observation that if we put $\gamma^t(s) = \gamma(t-s)$ then

$$\|\gamma^t\|_h^{\oplus} = O(\varepsilon). \tag{1.22}$$

As in I, in the following sections, unless we explicitly state otherwise, it is to be understood that the material description is used. No distinction is made between a material point and its location X in the reference configuration. When it is useful to emphasize that $F=F(t)$, F^t, and F_r^t depend not only on the time t but also on the material point X, we write $F(X,t)$, $F^t(X,\cdot)$, and $F_r^t(X,\cdot)$. Of course, for a given homogeneous material in a given motion, the stress-relaxation function G_t of (1.15) to (1.16) depends not only on t but also on X:

$$G_t(s)=\mathfrak{G}(F^t(X,\cdot);s). \tag{1.23}$$

According to whether it is the dependence on t or X, or both, that we wish to emphasize, we write G_t, G_X, or $G_{X,t}$, for this quantity. Similarly, for the numbers E_t and $\widetilde{E}_t$ of (1.8) and (1.9), we sometimes write E_X and $\widetilde{E}_X$, remembering, although this notation does not explicitly show it, that in general these moduli depend on both X and t.

2. General Properties of the Amplitude of Acceleration Waves

The *intrinsic velocity* of a wave is defined by

$$U(t)=\frac{d}{dt}Y(t) \tag{2.1}$$

where $Y(t)$, sometimes written Y_t, is the material point, labeled by its position in the reference configuration $\mathscr{R}$, at which the wave is to be found at time t. The *material trajectory* Σ of the wave is the set of ordered pairs (Y_t, t) with $-\infty<t<\infty$. If $f(X,t)$ has a jump discontinuity at $X=Y_t$ but is a continuous function of X and t jointly elsewhere, then we define the *jump in $f(X,t)$ across Σ at time t* by

$$[f](t)=f(Y_t^-,t)-f(Y_t^+,t). \tag{2.2}$$

It is important to our present subject to note that $[f]$ is a function of t alone. If we suppose that $\frac{\partial}{\partial X}f(X,t)$ and $\dot{f}(X,t)=\frac{\partial}{\partial t}f(X,t)$ are also continuous functions of X and t jointly except for jump discontinuities across Σ, then, on differentiating (2.2), we obtain

$$\frac{d}{dt}[f](t)=\frac{\partial f(Y_t^-,t)}{\partial X}\frac{dY_t}{dt}+\dot{f}(Y_t^-,t)-\left(\frac{\partial f(Y_t^+,t)}{\partial X}\frac{dY_t}{dt}+\dot{f}(Y_t^+,t)\right). \tag{2.3}$$

The equations (2.1)—(2.3) yield the following formula:*

$$\frac{d}{dt}[f]=[\dot{f}]+\left[\frac{\partial f}{\partial X}\right]U. \tag{2.4}$$

We say that $Y(t)$, $-\infty<t<\infty$, is an *acceleration wave* if the following two conditions hold.**

A-1: x, $\dot{x}$ and F are continuous functions of X and t jointly for all X and t, while $\ddot{x}$, $\frac{\partial F}{\partial X}$, $\dot{F}$ and also $\frac{\partial^2 F}{\partial X^2}$, $\frac{\partial \dot{F}}{\partial X}$, $\frac{\partial \ddot{x}}{\partial X}$, and $\dddot{x}$ have jump discontinuities across Σ but are continuous in X and t jointly everywhere else.

A-2:*** $F_r^t(X,\cdot)$ is a smooth function of X and t with respect to the norm $\|\cdot\|_h$.

* This formula is given in a more general context by TRUESDELL & TOUPIN [4, eq. (191.8)]. The history of equivalent formulae in the spatial description is given in footnote 2 to p. 504 of [4]. In the notation of THOMAS [5], $\frac{d}{dt}[f]$ is written $\frac{\delta}{\delta t}[f]$.

** These conditions A-1, A-2 are stronger than those laid down in § 5 of I.

*** In § 4 (Remark 4.1) we prove that this assumption is implied by A-1 whenever the wave is entering a homogeneous medium at rest.

More precisely A-2 means the following. For each X and t

(i) $F_r^t(X,\cdot)$ depends continuously on X and t with respect to $\|\cdot\|_h$ in the sense that $F_r^t(X,\cdot)$ belongs to $\mathscr{H}_h$ and $\lim_{\alpha,\beta\to 0}\|F_r^{t+\alpha}(X+\beta,\cdot)-F_r^t(X,\cdot)\|_h=0$.

(ii) There exist two functions in $\mathscr{H}_h$, denoted by $\frac{\partial}{\partial X}F_r^t(X,\cdot)$ and $\dot{F}_r^t(X,\cdot)$, such that

$$\lim_{\alpha\to 0}\left\|\frac{F_r^{t+\alpha}(X,\cdot)-F^t(X,\cdot)}{\alpha}-\dot{F}_r^t(X,\cdot)\right\|_h$$
$$=\lim_{\beta\to 0}\left\|\frac{F_r^t(X+\beta,\cdot)-F_r^t(X,\cdot)}{\beta}-\frac{\partial F_r^t(X,\cdot)}{\partial X}\right\|_h=0.$$

(iii) $\frac{\partial}{\partial X}F_r^t(X,\cdot)$ and $\dot{F}_r^t(X,\cdot)$ depend continuously on X and t with respect to $\|\cdot\|_h$.

We of course also assume that for each X and t, $F^t(X,\cdot)$ is interior to the domain $\mathscr{D}$ of $\mathfrak{T}$.

Every acceleration wave must obey the following compatibility conditions:*

$$[\ddot{x}]=-U[\dot{F}]=U^2\left[\frac{\partial F}{\partial X}\right]. \tag{2.5}$$

The jump in the acceleration is called the *amplitude* $a(t)$ of the acceleration wave:

$$a(t)=[\ddot{x}](t). \tag{2.6}$$

Putting $f=\ddot{x}$ and $f=\dot{F}$ in (2.4), we obtain

$$\frac{da}{dt}=\frac{d}{dt}[\ddot{x}]=[\dddot{x}]+[\ddot{F}]\,U, \tag{2.7}$$

$$\frac{d}{dt}[\dot{F}]=[\ddot{F}]+\left[\frac{\partial\dot{F}}{\partial X}\right]U. \tag{2.8}$$

Using (2.5) and (2.6), we can write (2.8) in the form

$$[\ddot{F}]\,U+\left[\frac{\partial\dot{F}}{\partial X}\right]U^2=U\frac{d}{dt}[\dot{F}]=-U\frac{d}{dt}\left(\frac{a}{U}\right)=-\frac{da}{dt}+\frac{a}{U}\frac{dU}{dt}, \tag{2.9}$$

and combining this with (2.7), we obtain

Remark 2.1. The amplitude of an acceleration wave obeys the equation

$$2\frac{da}{dt}-\frac{a}{U}\frac{dU}{dt}=[\dddot{x}]-U^2\left[\frac{\partial\dot{F}}{\partial X}\right]. \tag{2.10}$$

Note that equation (2.10) is purely kinematical. It involves neither the equations of motion nor the constitutive equation for T; in fact, in deriving (2.10) one uses only A-1.

In our present subject the *law of balance of momentum asserts* that for all times t and every pair of material points X_α, X_β

$$\frac{d}{dt}\int_{X_\alpha}^{X_\beta}\dot{x}(X,t)\,\varrho_R\,dX=\int_{X_\alpha}^{X_\beta}b(X,t)\,\varrho_R\,dX+T(X_\beta,t)-T(X_\alpha,t). \qquad (2.11)^{**}$$

---

* *Vid.*, equations (5.1a) and (5.2) of I or § 190 of [4].

** This equation (2.11) is the same as (3.8) of I, except for the new notation ϱ_R for the density in the reference configuration..

Here b, the body force per unit mass, is assumed to be assigned in advance in such a way that both $b(X, t)$ and $\dot{b}(X, t) = \frac{\partial}{\partial t} b(X, t)$ are continuous functions of X for all t; this is the only assumption we need about b.

Equation (2.11), when used in conjunction with our hypothesis on b and assumption A-1, implies that (a) for $X \neq Y_t$, T possesses a certain degree of smoothness and satisfies the equations

$$\frac{\partial T}{\partial X} + \varrho_R b = \varrho_R \ddot{x}, \tag{2.12a}$$

$$\frac{\partial^2 T}{\partial t \, \partial X} + \varrho_R \dot{b} = \varrho_R \dddot{x}; \tag{2.12b}$$

and (b) at $X = Y_t$ the jumps in $\frac{\partial T}{\partial X}$, $\ddot{x}$, and their time derivatives must obey the conditions

$$\left[\frac{\partial T}{\partial X}\right] = \varrho_R [\ddot{x}], \tag{2.12c}$$

$$\left[\frac{\partial^2 T}{\partial t \, \partial X}\right] = \varrho_R [\dddot{x}]. \tag{2.12d}$$

Equation (2.12d), when combined with Remark 2.1, yields

Remark 2.2. The amplitude of an acceleration wave obeys the equation

$$2 \frac{da}{dt} = \frac{a}{U} \frac{dU}{dt} + \frac{1}{\varrho_R} \left[\frac{\partial^2 T}{\partial t \, \partial X}\right] - U^2 \left[\frac{\partial \dot{F}}{\partial X}\right]. \tag{2.13}$$

It may be of interest to note that in deriving (2.13) we used only assumption A-1, a smoothness assumption on b, and balance of momentum.

Theorem 2 of I and (1.15) give us the following formula for the intrinsic velocity of an acceleration wave:

$$U^2(t) = G_{Y_t}(0)/\varrho_R; \tag{2.14}$$

here

$$G_{Y_t}(0) = \mathfrak{G}(F^t(Y_t, \cdot); 0) = D\mathfrak{T}(F^t(Y_t, \cdot)) \tag{2.15}$$

is the initial value of the stress-relaxation function G_{Y_t} corresponding to $F^t(Y_t, \cdot)$. Of course, we always have $\varrho_R > 0$. We further assume that $G_{Y_t}(0)$ is $\geqq 0$.

Now, using A-1, A-2, (1.10), some well-known chain rules, and the smoothness hypotheses on $\mathfrak{T}$ explained in the previous section (in particular that $\mathfrak{T}$ has continuous Fréchet-derivatives of orders 1 and 2, in the sense of (1.6), throughout its domain $\mathscr{D}$ which is an open set in $\mathscr{H}_h^{\oplus}$), we can write for $X \neq Y_t$

$$\frac{\partial}{\partial X} T(X, t) = D\mathfrak{T}(F^t(X, \cdot)) \frac{\partial F(X, t)}{\partial X} + \delta\mathfrak{T}\left(F^t(X, \cdot) \middle| \frac{\partial F_r^t(X, \cdot)}{\partial X}\right), \tag{2.16}$$

and, omitting some of the t's and most of the X's, but in otherwise the same notation,

$$\begin{aligned} \frac{\partial^2 T}{\partial t \, \partial X} = {} & D\mathfrak{T}(F^t) \frac{\partial \dot{F}}{\partial X} + D^2\mathfrak{T}(F^t) \frac{\partial F}{\partial X} \dot{F} + \delta D\mathfrak{T}(F^t | \dot{F}_r^t) \frac{\partial F}{\partial X} + \\ & + D\,\delta\mathfrak{T}\left(F^t \middle| \frac{\partial F_r^t}{\partial X}\right) \dot{F} + \delta^2\mathfrak{T}\left(F^t \middle| \frac{\partial F_r^t}{\partial X}, \dot{F}_r^t\right) + \frac{\partial}{\partial \tau} \delta\mathfrak{T}\left(F^t \middle| \frac{\partial F_r^\tau}{\partial X}\right)\bigg|_{\tau = t}. \end{aligned} \tag{2.17}$$

We now use equation (2.17) to evaluate the term $\left[\frac{\partial^2 T}{\partial t\,\partial X}\right]$ in (2.13). Since in A-2 we assume that $F_r^t(X,\cdot)$, as a point in $\mathscr{H}_h$, depends continuously on X and t, and in A-1 we assume that $F(X,t)$ is continuous in X and t, it follows that $F^t(X,\cdot)$, as a point in $\mathscr{H}_h^{\oplus}$, is a continuous function of X and t for all X and t. Hence, since $D\mathfrak{T}$ is a continuous functional over $\mathscr{H}_h^{\oplus}$, $[D\mathfrak{T}(F^t)]=0$, and we have

$$\left[D\mathfrak{T}(F^t)\frac{\partial \dot{F}}{\partial X}\right]=G_{Y_t}(0)\left[\frac{\partial \dot{F}}{\partial X}\right], \tag{2.18}$$

where $G_{Y_t}(0)$ is given by (2.15). Similarly, using (1.9) and the assumed continuity of $D^2\mathfrak{T}$, we get

$$\left[D^2\mathfrak{T}(F^t)\frac{\partial F}{\partial X}\dot{F}\right]=\widetilde{E}_{Y_t}\left[\frac{\partial F}{\partial X}\dot{F}\right], \tag{2.19}$$

where $\widetilde{E}_{Y_t}$ is the instantaneous second-order modulus corresponding to $F^t(Y_t,\cdot)$:

$$\widetilde{E}_{Y_t}=D^2\mathfrak{T}\left(F^t(Y_t,\cdot)\right). \tag{2.20}$$

A-2 asserts not only that $F_r^t(X,\cdot)$ in $\mathscr{H}_h$ is a differentiable function for all X and t but also that $\frac{\partial}{\partial X}F_r^t(X,\cdot)$ and $\dot{F}_r^t(X,\cdot)$ depend continuously on X and t; hence, the assumed continuity of $\delta^2\mathfrak{T}$, $D\delta\mathfrak{T}$, $\delta D\mathfrak{T}$ in all arguments permits us to write

$$\left[\delta D\mathfrak{T}(F^t|\dot{F}_r^t)\frac{\partial F}{\partial X}\right]=\left[\frac{\partial F}{\partial X}\right]I_{Y_t}, \tag{2.21}$$

$$\left[D\delta\mathfrak{T}\left(F^t\middle|\frac{\partial}{\partial X}F_r^t\right)\dot{F}\right]=[\dot{F}]\,J_{Y_t}, \tag{2.22}$$

$$\left[\delta^2\mathfrak{T}\left(F^t\middle|\frac{\partial}{\partial X}F_r^t,\dot{F}_r^t\right)\right]=0. \tag{2.23}$$

Here we have put

$$I_{Y_t}=\delta D\mathfrak{T}\left(F^t(Y_t,\cdot)|\dot{F}_r^t(Y_t,\cdot)\right), \tag{2.24}$$

$$J_{Y_t}=D\delta\mathfrak{T}\left(F^t(Y_t,\cdot)\middle|\frac{\partial}{\partial X}F_r^t(Y_t,\cdot)\right). \tag{2.25}$$

We now calculate the jump in the last term in (2.17) using (1.11), (1.12) and the assumptions made about G_t' immediately after (1.13). Clearly these assumptions and A-2 permit us to write

$$\begin{aligned}\frac{\partial}{\partial\tau}\delta\mathfrak{T}\left(F^t(X,\cdot)\middle|\frac{\partial}{\partial X}F_r^\tau(X,\cdot)\right)\Big|_{\tau=t}&=\frac{\partial}{\partial\tau}\int_0^\infty G'_{X,t}(s)\frac{\partial}{\partial X}F_r^\tau(X,s)\,ds\Big|_{\tau=t}\\&=\frac{\partial^2}{\partial\tau\,\partial Z}\mathscr{I}(X,t,Z,\tau)\Big|_{\substack{z=X\\ \tau=t}}\end{aligned} \tag{2.26}$$

where

$$\mathscr{I}(X,t,Z,\tau)=\int_0^\infty G'_{X,t}(s)\,F_r^\tau(Z,s)\,ds=\int_0^\infty G'_{X,t}(s)\,F(Z,\tau-s)\,ds. \tag{2.27}$$

By the assumed continuity in $\mathscr{H}_h$ of the dependence of $\frac{\partial}{\partial Z}F_r^\tau(Z,\cdot)$ on Z and τ, $\frac{\partial}{\partial Z}\mathscr{I}(X,t,Z,\tau)$ is jointly continuous in Z and τ. We want to show that

$$Z\neq Y_\tau\Rightarrow\frac{\partial^2}{\partial\tau\,\partial Z}\mathscr{I}(X,t,Z,\tau)=\frac{\partial^2}{\partial Z\,\partial\tau}\mathscr{I}(X,t,Z,\tau). \tag{2.28}$$

To do this it suffices to show that $\frac{\partial^2 \mathscr{I}}{\partial Z\,\partial \tau}$ exists and is jointly continuous in Z and τ for $Z \neq Y_\tau$. In view of A-2, we have

$$\begin{aligned}\frac{\partial}{\partial \tau}\mathscr{I}(X,t,Z,\tau) &= \int_0^\infty G'_{X,t}(s)\,\frac{\partial}{\partial \tau}F(Z,\tau-s)\,d\tau \\ &= -\int_0^\infty G'_{X,t}(s)\,\frac{\partial}{\partial s}F(Z,\tau-s)\,ds,\end{aligned} \tag{2.29}$$

and on integrating by parts we get

$$\frac{\partial}{\partial \tau}\mathscr{I}(X,t,Z,\tau) = -G'_{X,t}(s)\,F(Z,\tau-s)\Big|_0^\infty + \int_0^\infty G''_{X,t}(s)\,F(Z,\tau-s)\,ds. \tag{2.30}$$

Because of our hypotheses on $\mathfrak{T}$, the integral on the right in (1.11) exists for every function in $\mathscr{H}_h$. Since $G'_{X,t}(s) = K_{X,t}(s)\,h(s)^2$ with $K_{X,t}$ in $\mathscr{H}_h$ and, by A-2, $F_r^\tau(Z,\cdot)$ is in $\mathscr{H}_h$, $\int_0^\infty G'_{X,t}(s)\,F_r^\tau(Z,s)\,ds$ is finite. Hence, in view of the fact that $G'_{X,t}(s)\,F_r^t(Z,s)$ is a continuous function of s, it follows that

$$\lim_{s\to\infty} G'_{X,t}(s)\,F_r^\tau(Z,s) = 0 \tag{2.31}$$

for all Z, and (2.30) yields

$$\frac{\partial^2}{\partial Z\,\partial \tau}\mathscr{I}(X,t,Z,\tau) = G'_{X,t}(0)\,\frac{\partial}{\partial Z}F(Z,\tau) + \int_0^\infty G''_{X,t}(s)\,\frac{\partial}{\partial Z}F(Z,\tau-s)\,ds. \tag{2.32}$$

Equation (2.32) tells us that as long as we are not at a point where $Z = Y_\tau$ $\left(\textit{i.e.}, \text{ where } \frac{\partial}{\partial Z}F(Z,\tau) \text{ has a jump discontinuity}\right)$ $\frac{\partial^2}{\partial Z\,\partial \tau}\mathscr{I}$ is jointly continuous in Z and τ. Hence (2.28) holds, and by (2.26) and (2.32), for $X \neq Y_t$ we have

$$\begin{aligned}&\frac{\partial}{\partial \tau}\,\delta\mathfrak{T}\left(F^t(X,\cdot)\,\Big|\,\frac{\partial}{\partial X}F_r^\tau(X,\cdot)\right)\Big|_{\tau=t} \\ &\quad= G'_{X,t}(0)\,\frac{\partial}{\partial X}F(X,t) + \int_0^\infty G''_{X,t}(s)\,\frac{\partial}{\partial X}F(X,t-s)\,ds.\end{aligned} \tag{2.33}$$

It is now clear that A-1, A-2, and the assumed smoothness of $\mathfrak{G}$ insure that the last term on the right in (2.33) is continuous in X for all X and t. Hence

$$\left[\frac{\partial}{\partial \tau}\,\delta\mathfrak{T}\left(F^t\,\Big|\,\frac{\partial}{\partial X}F_r^\tau\right)\Big|_{\tau=t}\right] = G'_{Y_t}(0)\left[\frac{\partial F}{\partial X}\right], \tag{2.34}$$

where

$$G'_{Y_t}(0) = \frac{\partial}{\partial s}\,\mathfrak{G}\left(F^t(Y_t,\cdot);s\right)\big|_{s=0}. \tag{2.35}$$

Combining (2.17) with (2.18), (2.19), (2.21)—(2.23), and (2.34), we obtain

$$\left[\frac{\partial^2 T}{\partial t\,\partial X}\right] = G_{Y_t}(0)\left[\frac{\partial \dot F}{\partial X}\right] + \tilde{E}_{Y_t}\left[\frac{\partial F}{\partial X}\dot F\right] + \{I_{Y_t} + G'_{Y_t}(0)\}\left[\frac{\partial F}{\partial X}\right] + J_{Y_t}[\dot F], \tag{2.36}$$

which, upon use of (2.6), the compatibility conditions (2.5), and equation (2.14), becomes

$$\frac{1}{\varrho_R}\left[\frac{\partial^2 T}{\partial t\,\partial X}\right]=U^2\left[\frac{\partial \dot{F}}{\partial X}\right]-\frac{1}{\varrho_R U}\left\{J_{Y_t}-\frac{G'_{Y_t}(0)+I_{Y_t}}{U}\right\}a(t)+\frac{1}{\varrho_R}\tilde{E}_{Y_t}\left[\frac{\partial F}{\partial X}\dot{F}\right]. \tag{2.37}$$

Substitution of this result into (2.13) completes the proof of

Theorem 2.1. *The amplitude $a=a(t)$ of an acceleration wave obeys the relation*

$$\frac{da}{dt}=-\mu_t\,a+\beta_t\left[\frac{\partial F}{\partial X}\dot{F}\right], \tag{2.38}$$

where

$$\mu_t=-\frac{1}{2U}\frac{dU}{dt}+\frac{1}{2\varrho_R U}\left\{J_{Y_t}-\frac{G'_{Y_t}(0)+I_{Y_t}}{U}\right\}, \tag{2.39}$$

and

$$\beta_t=\frac{1}{2\varrho_R}\tilde{E}_{Y_t}; \tag{2.40}$$

$U=U(t)$ is the intrinsic velocity (2.1) *of the wave and satisfies* (2.14); *$G_{Y_t}(0)$ and $G'_{Y_t}(0)$ are, respectively, the initial value* (2.15) *and initial slope* (2.35) *of the stress-relaxation function corresponding to the history $F^t(Y_t,\cdot)$ at the wave; $\tilde{E}_{Y_t}$ is the instantaneous second-order modulus corresponding to $F^t(Y_t,\cdot)$; I_{Y_t} and J_{Y_t} are defined in* (2.24) *and* (2.25).

In the following two sections we shall integrate (2.38) to obtain explicit formulae for $a(t)$. To avoid discussion of reflection at edges we shall tacitly assume that the medium has infinite extent in the direction in which the wave is moving.

3. Application to Linear Viscoelasticity

The linear theory of viscoelasticity, defined by the constitutive assumption

$$T(t)=T_R+G(0)\,\gamma(t)+\int_0^\infty G'(s)\,\gamma(t-s)\,ds, \tag{3.1}$$

falls as a special case within the framework of the theory of simple materials with fading memory outlined in Section 1. In (3.1), $\gamma(t-s)=F(t-s)-1$ is the *strain* at time $t-s$ and G, with $G'(s)=\dfrac{dG(s)}{ds}$, is a material function, independent of F^t, called the *stress-relaxation function*; it is assumed that G is twice differentiable and that $G'h^{-1}$ is square integrable for some influence function h.

Writing (3.1) in the form

$$T(t)=\mathfrak{T}(F^t)=T_R+G(0)\,F(t)+\int_0^\infty G'(s)\,F_r^t(s)\,ds-G(\infty) \tag{3.2}$$

and looking at (1.7), one easily verifies that (3.1) implies

$$D\mathfrak{T}(F^t)=G(0), \tag{3.3a}$$

$$\delta\mathfrak{T}(F^t|g)=\int_0^\infty G'(s)\,g(s)\,ds, \tag{3.3b}$$

$$D^2\mathfrak{T}(F^t)=\delta^2\mathfrak{T}(F^t|g,l)=0. \tag{3.3c}$$

These equations tell us that in linear viscoelasticity the equations (2.38)—(2.40) of Theorem 1 must simplify considerably. Indeed, it follows from (3.3a) and (3.3b) that the function G_t defined for general materials by (1.11)—(1.15) is equal to the material function G and hence is independent of F^t; this tells us that $G_{Y_t}(0)$, $G'_{Y_t}(0)$, and U^2, being equal, respectively, to $G(0)$, $G'(0)$, and $G(0)/\varrho_R$, are here all independent of t. Furthermore, (3.3c), when placed beside (1.9), (2.24), and (2.25), yields zero for $\widetilde{E}_{Y_t}$, I_{Y_t} and J_{Y_t}. Thus, if (3.3) is assumed, then

$$\mu_t = -\frac{1}{2}\frac{G'(0)}{G(0)} = \text{const.}, \qquad \beta_t = 0, \tag{3.4}$$

(2.38) reduces to

$$\frac{da}{dt} = \frac{1}{2}\frac{G'(0)}{G(0)}\,a, \tag{3.5}$$

and Theorem 2.1 has the following corollary.

Theorem 3.1. *In the linear theory of viscoelasticity based on* (3.1), *the amplitude* a *of an acceleration wave has the following explicit dependence on the time:*

$$a(t) = a(0)\exp\left(\frac{G'(0)\,t}{2G(0)}\right). \tag{3.6}$$

Working in the theory of linear viscoelasticity, LEE & KANTER [*19*] and CHU [*6*] have exhibited explicit solutions of the dynamical equations showing shock waves. It is a trivial matter to construct from their results new solutions showing acceleration waves which, of course, obey (3.6). CHU pointed out that in his solutions the shock strength obeys (3.6). That the damping of shock waves and acceleration waves is of the same form in linear viscoelasticity is a consequence of the linearity of the field equations used in that theory.

4. A Wave Entering a Homogeneous Region at Rest

We now leave behind linear viscoelasticity and return to the general non-linear theory of simple materials with gradually fading memory.

We consider here an acceleration wave which since time $t=0$ has been propagating into a region which previously had been at rest in a fixed homogeneous configuration; *i.e.*, we suppose that $U(t)>0$ and that for $X \geqq Y(t)$ and $t \geqq 0$

$$x(X, t-s) = F_0 X + X_0, \qquad 0 \leqq s < \infty, \tag{4.1}$$

where F_0 and X_0 are constants. We shall see that for such a wave the differential equation (2.38) can be solved by quadrature for general materials. Before doing this, however, we record the following remark, the proof of which will be given at the end of this section.

Remark 4.1. Whenever (4.1) holds and $U(t)>0$, our hypothesis A-2 follows from our hypothesis A-1 for all $X>Y(0)$ and $t>0$.

In other words, in the present circumstances A-1 suffices to insure that $F_r^t(X,\cdot)$ is a smooth function of X and t with respect to $\|\cdot\|_h$, even when $X=Y_t$.

We now turn to the integration of (2.38). Let $F_0^\dagger$ be the constant history whose value is F_0; *i.e.*, let $F_0^\dagger$ be the *function* on $[0, \infty)$ defined by

$$F_0^\dagger(s) = F_0, \qquad 0 \leqq s < \infty. \tag{4.2}$$

It then follows from (4.1) that

$$F^t(X,\cdot)=F^t(Y_t,\cdot)=F_0^\dagger, \quad \text{for} \quad X>Y_t \quad \text{and} \quad t\geqq 0, \tag{4.3}$$

and hence that

$$G_{Y_t}(0)=G(0), \qquad G'_{Y_t}(0)=G'(0), \qquad \widetilde{E}_{Y_t}=\widetilde{E}, \quad \text{for} \quad t\geqq 0, \tag{4.4a}$$

where

$$G(0)=\mathfrak{G}(F_0^\dagger;0)=\text{const.}, \tag{4.4b}$$

$$G'(0)=\mathfrak{G}'(F_0^\dagger;0)=\text{const.}, \tag{4.4c}$$

and

$$\widetilde{E}=D^2\mathfrak{T}(F_0^\dagger)=\text{const.} \tag{4.4d}$$

In words: $G(0)$ and $G'(0)$ are, respectively, the initial value and the initial slope of the stress-relaxation function corresponding to the constant history $F_0^\dagger$; $\widetilde{E}$ is the instantaneous second-order modulus corresponding to $F_0^\dagger$. Note that by (2.14), $(4.4\text{a})_1$, and (4.4b), U is now independent of t:

$$U=\sqrt{\frac{G(0)}{\varrho_R}}. \tag{4.5}$$ *

The equation (4.1) also implies that

$$\dot{F}(X,t-s)=\frac{\partial}{\partial X}F(X,t-s)=0, \quad \text{whenever} \quad X>Y(t), \quad 0\leqq s<\infty, \quad \text{and} \quad t\geqq 0. \tag{4.6}$$

Since $\dot{F}_r^t(X,\cdot)$ and $\frac{\partial}{\partial X}F_r^t(X,\cdot)$ are, by A-2, both continuous functions of X, it follows from (4.3) that, for $t\geqq 0$, $\dot{F}_r^t(Y_t,\cdot)$ and $\frac{\partial}{\partial X}F_r^t(Y_t,\cdot)$ are both constant functions with value zero; thus, (2.24) and (2.25) here reduce to

$$I_{Y_t}=J_{Y_t}=0, \qquad t\geqq 0. \tag{4.7}$$

Obvious consequences of (4.3) are the relations

$$\dot{F}(Y_t^+,t)=\frac{\partial}{\partial X}F(Y_t^+,t)=0. \tag{4.8}$$

Therefore, by (2.2), we here have

$$\left[\frac{\partial F}{\partial X}\dot{F}\right]=\frac{\partial}{\partial X}F(Y_t^-,t)\,\dot{F}(Y_t^-,t)=\left[\frac{\partial F}{\partial X}\right][\dot{F}], \tag{4.9a}$$

and using (2.5) and (2.6) we can write

$$\left[\frac{\partial F}{\partial X}\dot{F}\right]=-U^{-3}a^2. \tag{4.9b}$$

It follows from this equation that (2.38) may be written

$$\frac{da}{dt}=-\mu a+\frac{\mu}{\lambda}a^2 \tag{4.10}$$

* Of course, ϱ_R is the density in the reference configuration, not the density ahead of wave.

where, by (4.3), (4.7), (2.39) and (2.40),

$$\mu = -\frac{G'(0)}{2G(0)} = \text{const.}, \qquad \lambda = \frac{G'(0)\,U}{\tilde{E}} = \text{const.} \tag{4.11}$$

Integrating (4.10), we obtain

Theorem 4.1. *Consider an acceleration wave propagating into an infinite medium which has always been at rest in a homogeneous configuration with deformation gradient F_0*; i.e., *suppose that $U(t) > 0$ and that* (4.1) *holds for $X \geqq Y(t)$ and $t \geqq 0$. Then, for $t \geqq 0$ the velocity U is a constant given by* (4.5), *and the amplitude a has the following explicit dependence on t:*

$$a(t) = \frac{\lambda}{\left(\frac{\lambda}{a(0)} - 1\right) e^{\mu t} + 1}. \tag{4.12}$$

For a given material μ and λ are constants depending on only F_0 as shown in (4.11) *and* (4.4).

In physical applications of the present theory we expect to have

$$G(0) > 0, \qquad G'(0) \leqq 0. \tag{4.13}$$

Indeed, $G(0) \geqq 0$ is a known consequence of the second law of thermodynamics.* We do not consider here the case $G(0) = 0$, for it yields $U = 0$. We cannot rule out with certainty the occurrence of a material with $G'(0) > 0$, but we know of none.**

When $G'(0) = 0$, (4.10) and (4.11) reduce to

$$\frac{da}{dt} = -\frac{\tilde{E}\,a^2}{2U\,G(0)}, \tag{4.14}$$

and one can calculate $a(t)$ by either integrating this equation or applying l'Hospital's rule to (4.12). The following result is obtained.

Remark 4.2. When one assumes that the hypotheses of Theorem 4.2 hold and, in addition, that $G'(0) = 0$, the amplitude a has the following dependence on t:

$$a(t) = \frac{a(0)}{1 + \nu\, a(0)\, t}, \qquad \nu = \frac{\tilde{E}}{2U\,G(0)}. \tag{4.15}$$

A special case of materials with $G'(0) = 0$ are perfectly elastic materials for which the equation (4.15) has been known for some time.*** Suppose that $\tilde{E} \neq 0$. Since $\varrho_R > 0$, (4.15) implies that if $a(0)$ has the same sign as $\tilde{E}$, then $a(t) \to 0$

* This has been shown by Coleman [*3*, § 8]. Both of the inequalities (4.13) have been discussed by Gurtin & Herrera [*7*].

** It follows from Theorem 4.2 of Gurtin & Herrera [*7*] that a material with $G'(0) > 0$ would not be "dissipative" in the sense of their definition (1.4).

*** For elastic fluids it was derived by T. Y. Thomas [*8*] in (1957). W. A. Green [*9*] has recently shown that an equation of the form (4.14) holds for all plane acceleration waves in homogeneously strained isotropic elastic materials. His results enable one to calculate ν given the stress-strain relation, the deformation gradient ahead of the wave, the direction of propagation, and the direction of the amplitude vector.

monotonically as $t \to \infty$; on the other hand, if $a(0)$ has the opposite sign to $\widetilde{E}$, then $|a(t)| \to \infty$ monotonically and in a finite time.

Another special case of Theorem 4.2 is obtained by assuming not that $G'(0)=0$, but rather that $\widetilde{E}=0$.

Remark 4.3. If in Theorem 4.1 one assumes that the instantaneous second-order modulus $\widetilde{E}$ is zero, then (4.12) reduces to

$$\frac{a(t)}{a(0)} = \exp\left(\frac{G'(0)\,t}{2G(0)}\right). \tag{4.16}$$

Equation (4.16) is of the same form as equation (3.6), derived, for more general kinematical situations, in the linear theory of viscoelasticity. Of course, in (4.16) $G'(0)$ and $G(0)$ may depend on F_0.

We may note that if we put $\varepsilon=a(0)/\lambda$, with λ given by (4.11), and assume $G(0)>0$, $G'(0)<0$, then, even if $\widetilde{E}\neq 0$, whenever ε is small in magnitude (4.12) yields

$$\frac{a(t)}{a(0)} = \exp\left\{\frac{G'(0)}{2G(0)}\,t\right\}(1+O(\varepsilon))$$

where $O(\varepsilon)$ is uniform in t; *i.e.*, for small ε, $O(\varepsilon)<M\varepsilon$ for some M independent of t.

Supposing that $G'(0)<0$, we observe that (4.16) implies that $a(t)\to 0$ as $t\to\infty$, regardless of the sign of $a(0)$. The gap between this observation and that following from the classical equation (4.15) for nonlinear but elastic materials is filled by the following theorem, which is valid for all simple materials with gradually fading memory.

Let us call the number $|\lambda|$, defined by (4.11), the *critical amplitude*.

Theorem 4.2. *Let the hypothesis of Theorem 4.1 hold, and suppose that*

$$G(0)>0, \qquad G'(0)<0, \qquad \widetilde{E}\neq 0. \tag{4.17}$$

If either

(i) $|a(0)|$ *is less than the critical amplitude, or*

(ii) $\operatorname{sgn} a(0) = \operatorname{sgn} \widetilde{E}$,

then $a(t)\to 0$ *monotonically as* $t\to\infty$. *If* $a(0)=\lambda$, *then* $a(t)\equiv a(0)$. *On the other hand, if both*

(i*) $|a(0)|$ *is greater than the critical amplitude, and*

(ii*) $\operatorname{sgn} a(0) = -\operatorname{sgn} \widetilde{E}$,

then $|a(t)|\to\infty$ *monotonically and in a finite time* t_∞ *given by*

$$t_\infty = -\frac{1}{\mu}\ln\left(1-\frac{\lambda}{a(0)}\right). \tag{4.18}$$

We omit the details of the proof; it involves no more than an elementary analysis of the function on the right in (4.12).

This last theorem asserts that, assuming (4.17), if the amplitude is sufficiently small or if the amplitude has the same sign as the instantaneous second-order modulus, then an acceleration wave obeying (4.1) is gradually damped out; if, however, the amplitude exceeds the critical amplitude and has its sign opposite to that of the instantaneous second-order modulus, then the wave blows up, *i.e.*, achieves an infinite amplitude in a finite time, which may mean that a shock wave is produced. We note that *the presence of internal damping*, manifested by a strictly negative value of $G'(0)$, *does not always imply that a singular surface moving into a homogeneous region must be damped out.*

We now give a *proof of Remark* 4.1. Let X and t, with $t>0$ and $X>Y(0)$, be chosen arbitrarily and then held fixed throughout our discussion. It suffices to show that

$$\int_0^\infty |F^t(X,s)\,h(s)|^2\,ds<+\infty,\qquad \int_0^\infty \left|\frac{\partial}{\partial X}F^t(X,s)\,h(s)\right|^2 ds<+\infty,$$
$$\int_0^\infty |\dot{F}^t(X,s)\,h(s)|^2\,ds<+\infty, \tag{4.19}$$

$$\lim_{\alpha,\beta\to 0}\int_0^\infty |F^{t+\alpha}(X+\beta,s)-F^t(X,s)|^2\,h(s)^2\,ds=0, \tag{4.20a}$$

$$\lim_{\alpha,\beta\to 0}\int_0^\infty \left|\frac{\partial}{\partial X}F^{t+\alpha}(X+\beta,s)-\frac{\partial}{\partial X}F^t(X,s)\right|^2 h(s)^2\,ds=0, \tag{4.20b}$$

$$\lim_{\alpha,\beta\to 0}\int_0^\infty |\dot{F}^{t+\alpha}(X+\beta,s)-\dot{F}^t(X,s)|^2\,h(s)^2\,ds=0, \tag{4.20c}$$

$$\lim_{\beta\to 0}\int_0^\infty \left|\frac{F^t(X+\beta,s)-F^t(X,s)}{\beta}-\frac{\partial F^t(X,s)}{\partial X}\right|^2 h(s)^2\,ds=0, \tag{4.21a}$$

and

$$\lim_{\alpha\to 0}\int_0^\infty \left|\frac{F^{t+\alpha}(X,s)-F^t(X,s)}{\alpha}-\dot{F}^t(X,s)\right|^2 h(s)^2\,ds=0. \tag{4.21b}$$

Since h is square integrable, if $X>Y(t)$, then (4.19), (4.20), and (4.21) follow at once, since $F^t(Z,\cdot)$ equals the constant function $F_0^\dagger$ for all (Z,t) in a sufficiently small neighborhood of (X,t). Therefore, we assume $X\leqq Y(t)$. Let Ω and Ω^+ be the sets

$$\Omega=\{(Z,t):\ X-\lambda\leqq Z\leqq X+\lambda,\ -\infty<t\leqq t+\lambda\}, \tag{4.22a}$$
$$\Omega^+=\{(Z,t):\ (Z,t)\in\Omega,\ Z\leqq Y(t)\}, \tag{4.22b}$$

where $\lambda>0$ is chosen such that $Y(0)<X-\lambda$. Clearly Ω^+ is a compact set, and a portion of its boundary coincides with the material trajectory Σ of the wave. However, Σ and the interior of Ω^+ are disjoint. Hence A-1 implies the existence of a positive number M which bounds $|F(Z,t)|$, $|\dot{F}(Z,t)|$, and $\left|\frac{\partial F(Z,t)}{\partial Z}\right|$ on Ω^+. Moreover, on $\Omega-\Omega^+$, $F(Z,t)=F_0$ (a constant). Thus we have the inequalities

$$|F(Z,t)|<M,\qquad |\dot{F}(Z,t)|<M,\qquad \left|\frac{\partial F(Z,t)}{\partial Z}\right|<M, \tag{4.23}$$

for every $(Z,t)\in\Omega$. An immediate consequence of (4.23) is (4.19).

Let s^* be the unique real number with the property that $X=Y_{t-s^*}$; s^* is unique because $U(\cdot)>0$. The three integrands in (4.20) converge to zero as $\alpha,\beta\to 0$ for every $s\neq s^*$. Moreover, by (4.23), each of the integrands in (4.20) is bounded, for all sufficiently small α and β, by the integrable function $4M^2h(s)^2$. Therefore Lebesgue's dominated convergence theorem implies (4.20).

Next we have, for sufficiently small α and β, the inequalities:

$$|F^{t+\alpha}(X,s)-F^t(X,s)|<2M\alpha,\qquad |F^t(X+\beta,s)-F^t(X,s)|<2M\beta. \tag{4.24}$$

These inequalities are implied by A-1, (4.23), and the following two facts: (i) Since $U(\cdot)>0$, any line $Z=$constant which lies in Ω, or any line $t=$constant, can intersect Σ in at most a single point. (ii) For any continuous function f on a closed interval

$[a, b]$ whose derivative f' is also continuous on $[a, b]$, except possibly for a jump discontinuity at a single point, $\sup_{[a,b]} f \leqq 2(b-a) \sup_{[a,b]} f'$. We are now in a position to bound the integrands in (4.21). Indeed, by (4.23) and (4.24) these integrands are each bounded by the integrable function $9M^2 h(s)^2$. Also these integrands converge to zero as $\alpha, \beta \to 0$ for every $s \neq s^*$. Hence, a second application of Lebesgue's theorem implies (4.21). This completes the proof.

5. Application to Longitudinal Waves

Our present one-dimensional theory may be applied to the study of longitudinal waves proceeding along the axis of a cylinder composed of an isotropic, or transversely isotropic, simple material.* In this application X denotes the distance along the cylindrical axis, and $F(X, t)$, which we take to be positive, determines the density $\varrho(X, t)$ as follows:**

$$\frac{1}{\varrho(X,t)} = F(X,t)\,\frac{1}{\varrho_R}. \tag{5.1}$$

Hence, in a longitudinal acceleration wave $\varrho(X, t)$ is continuous for all X and t, while $\dot{\varrho}(X, t)$ and $\frac{\partial}{\partial X}\varrho(X, t)$ have jump discontinuities across Σ but are continuous in X and t jointly everywhere else. We call the wave *compressive* at time t if

$$U(t)\,[\dot{\varrho}]\,(t) > 0 \tag{5.2a}$$

and *expansive* at t if

$$U(t)\,[\dot{\varrho}]\,(t) < 0. \tag{5.2b}$$

Thus the wave is compressive or expansive according to whether its passage through a point results in an increase or decrease in $\dot{\varrho}$ at the point.

Combining the equations (5.1), $(2.5)_1$, and (2.6), we obtain

$$a = \varrho_R\, \varrho^{-2}\, U[\dot{\varrho}], \tag{5.3}$$

from which follows

Remark 5.1. A longitudinal acceleration wave is compressive at t if $a(t) > 0$ and expansive if $a(t) < 0$.

Further, from $(4.11)_2$, (5.1), and (5.3), we see that

$$a(0) = \lambda \Leftrightarrow [\dot{\varrho}]\,(0) = \lambda_*, \tag{5.4}$$

where

$$\lambda_* = \frac{G'(0)\,\varrho_R}{\tilde{E}\, F_0^2}. \tag{5.5}$$

We call $|\lambda_*|$ the *critical jump in the density rate.* This name is justified by the following remark which follows immediately from Remark 5.1 and Theorem 4.1.

---

* The theoretical basis of this statement and some of the experimental difficulties involved in achieving purely longitudinal motions in actual three-dimensional specimens are discussed in § 6 of I. Our present *longitudinal waves* are called *plane longitudinal waves* by GREEN [9] and *plane dilatational waves* by BLAND [10] in their studies of waves in elastic materials.

** See equation (6.19) of I.

Remark 5.2. Consider a longitudinal acceleration wave obeying the hypothesis of Theorem 4.1 and propagating into a material obeying (4.17). *If either*

(i) $|[\dot{\varrho}](0)| < |\lambda_*|$, or

(ii) the wave is initially compressive with $\tilde{E} > 0$, or

(iii) the wave is initially expansive with $\tilde{E} < 0$,

then $[\dot{\varrho}](t) \to 0$ monotonically as $t \to \infty$. *If* $[\dot{\varrho}](0) = \lambda_*$, *then* $[\dot{\varrho}](t) \equiv [\dot{\varrho}](0)$. But, *if*

(i*) $|[\dot{\varrho}](0)| > |\lambda_*|$, *and either*

(ii*) the wave is initially compressive with $\tilde{E} < 0$, or

(iii*) the wave is initially expansive with $\tilde{E} > 0$,

then $|[\dot{\varrho}(t)| \to \infty$ monotonically and in the finite time

$$t_\infty = \frac{2G(0)}{G'(0)} \ln\left(1 - \frac{\lambda_*}{[\dot{\varrho}](0)}\right). \tag{5.6}$$

A qualitative, experimental verification of Remark 5.2 is obtained for a gas whenever one generates a shock wave from a sound impulse in a shock tube. The motion of a gas in a shock tube can be assumed to be longitudinal. For most gases, including ideal gases, $\tilde{E} < 0$; therefore, Remark 5.2 tells us that an acceleration wave in a tube should grow into a shock wave only if it is compressive, and this is precisely what is observed. The remark also tells us that the shock wave should not form unless $|[\dot{\varrho}]|$ is initially greater than the critical jump in the density-rate given by (5.5). We do not know whether this critical jump has been measured precisely, but surely it exists; otherwise every sudden burst of sound, no matter how faint, as it moves down a long tube of air would eventually grow into a shock.

6. Application to Shear Waves

Consider a three-dimensional simple incompressible isotropic body $\mathscr{B}$ with fading memory, and let x^1, x^2, x^3 be the Cartesian coordinates at time t of that material point which has the coordinates X^1, X^2, X^3 when $\mathscr{B}$ is in a fixed homogeneous reference configuration $\mathscr{R}$ with density ϱ_R. We say that the motion of $\mathscr{B}$ is a *pure shear* if for all t

$$\begin{aligned} x^1 &= X^1, \\ x^2 &= X^2 + \lambda(X^1, t), \\ x^3 &= X^3. \end{aligned} \tag{6.1}$$

We call a three-dimensional wave for which the motion has the form (6.1) for all t a *shear-wave.** Let us denote by T^{ij} the Cartesian components of the stress and suppose that the body forces arise from a potential and have no component along the x^2-axis. In I, assuming that $\mathscr{R}$ is an undistorted** reference configuration of $\mathscr{B}$, it was shown that if one puts T equal to the shearing stress T^{12},

$$T = T^{12} = T^{21}, \tag{6.2}$$

---

* It may also be called a *plane transverse wave with constant polarization.*

** We use the word *undistorted* in the sense made precise by Noll [2].

and further puts

$$X = X^1, \tag{6.3 a}$$

$$x = x^2 - X^2 + X^1 = \lambda(X^1, t) + X^1 = x(X, t), \tag{6.3 b}$$

$$F(X, t) = \frac{\partial}{\partial X} x(X, t) = \frac{\partial}{\partial X^1} \lambda(X^1, t) + 1, \tag{6.3 c}$$

then T is given, as in (1.1), by a smooth material functional $\mathfrak{T}$ of F^t, and a necessary and sufficient condition that the motion (6.1) be compatible with the three-dimensional integral form of the law of balance of momentum is that the one-dimensional integral form (2.11) holds with, of course, $b=0$.* Hence, when we make the identifications (6.2) and (6.3), acceleration waves involving only the pure shear of a three-dimensional incompressible isotropic material with fading memory are covered by the one-dimensional theory developed here and in I. However, in the application of the one-dimensional theory to pure shears, the assumed isotropy of the three-dimensional body implies that T be given by an odd function of the strain history $\gamma^t = F^t - 1^\dagger$, *i.e.*, that $\mathfrak{T}$ obey the identity**

$$\mathfrak{T}(1^\dagger + \gamma^t) = -\mathfrak{T}(1^\dagger - \gamma^t) \tag{6.4}$$

for all functions γ^t on $[0, \infty)$ such that $F^t = 1^\dagger + \gamma^t$ is in $\mathscr{D}$. It follows that $\mathfrak{T}(1_r^\dagger; 1+\gamma)$ is an odd function of the real variable $\gamma = \gamma(t)$. Hence, by (1.7b),

$$D^2\mathfrak{T}(1^\dagger) = \frac{\partial^2}{\partial \gamma^2} \mathfrak{T}(1_r^\dagger; 1+\gamma)\Big|_{\gamma=0} = 0. \qquad (6.5)^{***}$$

The conclusion that our one-dimensional theory can be applied to shearing motions of isotropic media remains valid not only if the reference configuration $\mathscr{R}$ is undistorted but also even if $\mathscr{R}$ differs by a stretch from an undistorted reference configuration, provided that the stretch is homogeneous and that its principal directions of strain lie along the coordinate axes x^1, x^2, x^3; i.e., provided only that the coordinates X^1, X^2, X^3 of a material point in the configuration $\mathscr{R}$ are related as follows to the coordinates $\overline{X}^1, \overline{X}^2, \overline{X}^3$ of the point when $\mathscr{B}$ is in an undistorted reference configuration $\overline{\mathscr{R}}$:

$$\begin{aligned} X^1 &= A^1 + \alpha^1 \overline{X}^1, \\ X^2 &= A^2 + \alpha^2 \overline{X}^2, \\ X^3 &= A^3 + \alpha^3 \overline{X}^3. \end{aligned} \tag{6.6}$$

(By the incompressibility of the body, $\alpha_1\alpha_2\alpha_3 = 1$.) *In this more general application, the functional* $\mathfrak{T}$ *of* (1.1), *though it depends on the principal stretch ratios* α_i, *still obeys the identity* (6.4)**** *and hence has the property* (6.5).

* See equations (6.9)—(6.20) of I and the discussion thereafter.

** This is equation $(6.28)_1$ of I, which identity was derived by COLEMAN & NOLL [*11*] for incompressible fluids. A proof that it holds for incompressible isotropic materials in general will be given in [*12*].

*** Actually (6.4) implies the stronger result that $\delta^{\oplus 2}\mathfrak{T}(1^\dagger | f, g) = 0$ for all functions f and g in $\mathscr{H}_h^{\oplus}$.

**** A formal proof of this intuitively evident statement will be given in [*12*]. We may note that the statement remains true if the material, instead of being isotropic, is only transversely isotropic with its axis of symmetry in the x^3-direction.

We may now apply our Theorem 4.1 to a shear wave propagating along a principal axis of strain into a statically and homogeneously deformed incompressible material. In such an application for $X > Y(t)$ and $t \geqq 0$ we have, in (6.1), $x^2 = X^2$, and hence, by (6.3), $F_0 = 1$ in (4.2)—(4.4). The equations (4.4) become

$$G(0) = \mathfrak{G}(1^\dagger; 0), \tag{6.7a}$$

$$G'(0) = \mathfrak{G}'(1^\dagger; 0), \tag{6.7b}$$

$$\widetilde{E} = D^2 \mathfrak{T}(1^\dagger). \tag{6.7c}$$

Since our reference configuration $\mathscr{R}$ and hence $\mathfrak{T}$ here depend on the principal stretch ratios $\alpha_1, \alpha_2, \alpha_3$ for the region ahead of the wave, so also do the moduli $G(0)$ and $G'(0)$. More importantly, we see that (6.6) and (6.7c) imply that

$$\widetilde{E} = 0 \tag{6.8}$$

and hence that (4.12) reduces to (4.16). This proves

Remark 6.1. At a shear acceleration wave propagating along a principal axis of strain into an isotropic material which has always been at rest in a fixed homogeneous configuration, the instantaneous second-order modulus $\widetilde{E}$ must be zero. It follows from Theorem 4.1 that the amplitude of such a wave obeys equation (4.16) with G the shear-stress relaxation function for the strained material ahead of the wave. If we assume, as is reasonable, that $G(0) > 0$ and that $G'(0) < 0$, then we are forced to the conclusion that the amplitude of the wave will decay to zero as $t \to \infty$.

It goes without saying that all the conclusions of the above remark hold for a shear wave entering an isotropic material which was previously always at rest in an undistorted configuration. We emphasize that equation (4.16), though it has the same form as equation (3.6) of linear viscoelasticity, has here been derived as an exact result, valid for certain shear waves regardless of the magnitude of the amplitude $a(0)$.

If in Remark 6.1 we replace the assumption $G'(0) < 0$ with the appropriate assumption for perfectly elastic materials, *i.e.*, $G'(0) = 0$, then we conclude, as did GREEN [6] in his study of elastic waves, that the amplitude of the shear wave remains constant in time.

7. Relation of Acceleration Waves to Infinitesimal Progressive Waves

For a one-dimensional motion under zero body force, use of the displacement function,

$$w(X, t) \equiv x(X, t) - X, \tag{7.1}$$

permits one to write the dynamical equation (2.12a) in the form

$$\varrho_R \ddot{w} = \frac{\partial}{\partial X} \mathfrak{T}(F^t) = \delta^{\oplus} \mathfrak{T}\left(F^t \,\Big|\, \frac{\partial F^t}{\partial X}\right) = \delta^{\oplus} \mathfrak{T}\left(1^\dagger \,\Big|\, \frac{\partial^2 w^t}{\partial X^2}\right) + \mathfrak{R}, \tag{7.2}$$

where w^t is a function depending on X and defined by

$$w^t(s) = w(X, t - s), \qquad 0 \leqq s < \infty, \tag{7.3}$$

and for fixed $\frac{\partial^2 w^t}{\partial X^2}$,

$$\mathfrak{R} = o(1) \tag{7.4}$$

in $\left\|\frac{\partial w^t}{\partial X}\right\|_h^{\oplus}$. If we let G be the stress relaxation function corresponding to the history $F^t = 1^\dagger$, *i.e.*, the relaxation function of that linearly viscoelastic material which approximates the general simple material in small deformations, then, by (1.17), (7.2) may be rewritten as

$$\varrho_R \ddot{w}(X,t) = G(0) \frac{\partial^2}{\partial X^2} w(X,t) + \int_0^\infty G'(s) \frac{\partial^2}{\partial X^2} w(X, t-s)\, ds + \mathfrak{R}. \tag{7.5}$$

The *one-dimensional dynamical theory of linear viscoelasticity* is obtained by neglecting the term $\mathfrak{R}$ in (7.5). The very difficult question of when $\mathfrak{R}$ may be neglected without seriously affecting the character of solutions of this functional-partial differential equation is at present unanswered. Instead of attempting a discussion of this question, we now consider the limitations which the dynamical theory of linear viscoelasticity places on a motion for which

$$\begin{gathered} w(X,t) = A\, e^{-\alpha X} \cos\omega\left(t - \frac{X}{c}\right) = A \operatorname{Re}\left\{e^{-\left(\alpha + \frac{i\omega}{c}\right)X} e^{i\omega t}\right\}, \\ -\infty < t < \infty, \qquad 0 \leqq X < \infty, \end{gathered} \tag{7.6}$$

with $\omega > 0$, $\alpha > 0$, $c > 0$ and A all real numbers. Such a motion is called a *damped infinitesimal sinusoidal progressive wave* with *frequency* ω, *attentuation* α, *speed* c and *amplitude* A.

The progressive wave (7.6) is called "infinitesimal" because in a general simple material it can be dynamically admissible only if $\mathfrak{R}$ is neglected, in which case (7.6) clearly satisfies (7.5) if and only if

$$-\varrho_R \omega^2 = \left(\alpha + \frac{i\omega}{c}\right)^2 \{G(0) + \overline{G'}(\omega)\} \tag{7.7a}$$

where

$$\overline{G'}(\omega) = \int_0^\infty G'(s)\, e^{-i\omega s}\, ds. \tag{7.7b}$$

For each ω there is exactly one pair of values $c = c(\omega) > 0$, $\alpha = \alpha(\omega) > 0$, satisfying (7.7):*

$$c^2(\omega) = \frac{|G(0) + \overline{G'}(\omega)|}{\varrho_R} \sec^2 \frac{\vartheta(\omega)}{2}, \tag{7.8a}$$

$$\alpha(\omega) = \frac{\omega}{c(\omega)} \tan \frac{\vartheta(\omega)}{2}, \tag{7.8b}$$

where $0 \leqq \vartheta < \frac{\pi}{2}$ and

$$\tan\vartheta(\omega) = \frac{\operatorname{Im}\{G(0) + \overline{G'}(\omega)\}}{\operatorname{Re}\{G(0) + \overline{G'}(\omega)\}}. \tag{7.8c}$$

---

* The equations (7.8) appear to be well known. See, for example, the recent survey by HUNTER [*20*].

We now assume that $G'(s)$ and $G''(s) = \frac{d}{ds} G'(s)$ are both continuous and absolutely integrable on $[0, \infty)$. It then follows from (7.7b) and well-known theorems on Fourier transforms that

$$\lim_{\omega\to\infty} \overline{G'}(\omega) = 0 \tag{7.9a}$$

and

$$\lim_{\omega\to\infty} \omega\, \overline{G'}(\omega) = -i\, G'(0). \tag{7.9b}$$

Using these formulae, one can easily show that the equations (7.8) imply that the limits,

$$c_\infty = \lim_{\omega\to\infty} c(\omega), \tag{7.10a}$$

and

$$\alpha_\infty = \lim_{\omega\to\infty} \alpha(\omega), \tag{7.10b}$$

which are called, respectively, the *ultrasonic speed* and the *ultrasonic attenuation*, exist and are given by

$$c_\infty = \sqrt{\frac{G(0)}{\varrho_R}}, \tag{7.11a}$$

$$c_\infty\, \alpha_\infty = -\frac{G'(0)}{2G(0)}. \tag{7.11b}$$

Setting this result alongside Theorem 4.1, we read off

Remark 7.1. Let $U(t) > 0$ be the intrinsic velocity and $a(t)$ the amplitude of an acceleration wave which since $t = 0$ has been propagating into an infinite homogeneous body that previously was always at rest in its reference configuration. Then, for $t \geqq 0$

$$U(t) = c_\infty, \tag{7.12}$$

and $a(t)$ is given by (4.12) with

$$\mu = c_\infty \alpha_\infty, \tag{7.13a}$$

$$\lambda = \frac{-2c_\infty^4\, \alpha_\infty\, \varrho_R}{\widetilde{E}}, \tag{7.13b}$$

where c_∞ and α_∞ are the ultrasonic speed and ultrasonic attenuation for damped infinitesimal sinusoidal progressive waves in the same body.

The equation (7.12) gives to materials with memory an extension of TRUESDELL'S* "second theorem of equivalence" for the velocities of acceleration waves and plane infinitesimal progressive waves in perfectly elastic materials subject to homogeneous strain.

We may note that if $\widetilde{E}$ happens to be zero, as it is for the shear waves covered by Remark 6.1, then (7.13) and (7.12), when combined with (4.16), tell us that the ultrasonic attenuation α_∞ determines the amplitude a of the acceleration wave of Remark 7.1 as follows:

$$a(t) = a(0)\, e^{-c_\infty \alpha_\infty t}, \tag{7.14a}$$

or

$$\hat{a}(Y_t) = \hat{a}(0)\, e^{-\alpha_\infty Y_t}, \tag{7.14b}$$

* [*13*, § 4].

where the function $\hat{a}$ is defined to be such that $\hat{a}(Y_t)=a(t)$ and Y_t is taken to be 0 at $t=0$.

Writing (7.6) in the form

$$w(X,t)=\mathcal{A}_\omega(X)\cos\omega\left(t-\frac{X}{c(\omega)}\right),\qquad \mathcal{A}_\omega(X)=A\,e^{-\alpha(\omega)X},\tag{7.15}$$

we may consider the following *similarity relations between damped infinitesimal progressive waves and acceleration waves entering a body in a given static homogeneous configuration:*

$$U=\lim_{\omega\to\infty}c(\omega),\tag{7.16}$$

$$\frac{\hat{a}(Y_t)}{\hat{a}(0)}=\lim_{\omega\to\infty}\frac{\mathcal{A}_\omega(Y_t)}{\mathcal{A}_\omega(0)}.\tag{7.17}$$

Unfortunately, Remark 7.1 tells us that only the first of these equations is true in general; the second holds only if $\tilde{E}=0$. When $\tilde{E}\neq 0$, equation (7.17) must be replaced by the more complicated relation obtained by inserting (7.13) into (4.12).

8. On the Physical Meaning of the Modulus $\tilde{E}_t$

We saw in Section 4 that a necessary and sufficient condition for growth of the amplitude a of a one-dimensional acceleration wave propagating into a homogeneous region at rest is that a be initially of the same sign as $\tilde{E}$ and larger in magnitude than a critical amplitude $|\lambda|=\left|\frac{G'(0)\,U}{\tilde{E}}\right|=\left|\frac{G(0)^{\frac{1}{2}}G'(0)}{\varrho_R^{\frac{1}{2}}\tilde{E}}\right|$. If this condition is met, a will become infinite in the time interval t_∞ given by (4.18). Looking at (4.4) and Remark 1.1, one sees that $G(0)$ and $G'(0)$ are here just the initial value and initial slope of the stress-relaxation function,

$$G(\cdot)=\mathfrak{G}(F_0^{\dagger};\cdot),\tag{8.1}$$

of that linearly viscoelastic material which approximates our general nonlinear material in small deformations from a homogeneous configuration. Hence, in view of the large literature on methods of measuring stress-relaxation functions of linear viscoelasticity,* the problem of knowing what physical properties determine the sign and magnitude of λ is really one of understanding the physical significance of the parameter $\tilde{E}$, which does not occur in linear theories. In this section and the next we shall relate $\tilde{E}$ to quantities such as stress, strain-rate, wave-velocity, *etc.*, which are well understood. Since, however, we do not exhibit dynamically admissible motions giving the strain histories we study, we cannot assert that the following discussions give practical means of measuring $\tilde{E}$.

Linear Approximations to Jump Continuations

Let Ξ be a number and F^t an arbitrary history. The function F^{t*} defined by

$$F^{t*}(s)=\begin{cases}F^t(0)+\Xi, & s=0,\\ F^t(s), & 0<s<\infty\end{cases}\tag{8.2}$$

* See, for example, the books of GROSS [14] and FERRY [15].

is called the *jump continuation* of F^t *with jump* Ξ.* In terms more physically suggestive than precise, we can say that F^{t*} differs from F^t by only a superimposed strain-impulse occurring at time t and having intensity Ξ. Let us keep F^t fixed and regard the stress T^* corresponding to F^{t*} as a function of Ξ:

$$T^*(\Xi) \stackrel{\text{def}}{=} \mathfrak{T}(F^{t*}). \tag{8.3}$$

The instantaneous tangent modulus $E_t = G_t(0)$ and the instantaneous second-order modulus $\widetilde{E}_t$ corresponding to F^t are defined in equations (1.8) and (1.9); clearly, these equations are equivalent to the assertions that

$$E_t = \frac{d}{d\,\Xi} T^*(\Xi)\Big|_{\Xi=0}, \tag{8.4a}$$

$$\widetilde{E}_t = \frac{d^2}{d\,\Xi^2} T^*(\Xi)\Big|_{\Xi=0}; \tag{8.4b}$$

i.e., E_t and $\widetilde{E}_t$ are the first and second rates of change of stress with respect to the jump in strain in a jump continuation.

Fortunately, a stress-strain function $T^*(\cdot)$ measured suddenly at a given instant t using strain-impulses can be regarded as a limit of stress-strain functions obtained by measuring the stress at times after t using continuations of F^t which are continuous at t but involve very high rates of strain after t. To make this precise, suppose that we are given a material at time t with a certain history F^t and that from t to some other time $t+\delta$, $\delta>0$, we deform the material at a rate $\frac{1}{\delta}\Xi$. Then, the history of the material at time $t+\delta$ will be given by

$$F^{t+\delta,\Xi}(s) = \begin{cases} F^t(0) + \dfrac{\delta - s}{\delta}\,\Xi, & 0 \leqq s < \delta, \\ F^t(s-\delta), & \delta \leqq s < \infty. \end{cases} \tag{8.5}$$

For a given history F^t, the stress at time $t+\delta$ corresponding to such a *linear continuation* $F^{t+\delta,\Xi}$ will depend on both δ and Ξ:

$$T^{(\delta)}(\Xi) = \mathfrak{T}(F^{t+\delta,\Xi}). \tag{8.6}$$

Note that the smaller is δ, the closer $t+\delta$ is to t but also the faster is the rate $\frac{1}{\delta}\Xi$. It follows from (1.4) and (1.5) that as $\delta\to 0$ the linear continuation (8.5) converges, in $\mathscr{H}_h^{\oplus}$, to the jump continuation defined in (8.2); *i.e.*, for any F^t in $\mathscr{H}_h^{\oplus}$

$$\lim_{\delta\to 0} \|F^{t+\delta,\Xi} - F^{t*}\|_h^{\oplus} = 0. \tag{8.7**}$$

Hence, by our smoothness hypotheses on $\mathfrak{T}$, we have

$$\lim_{\delta\to 0} T^{(\delta)}(\Xi) = T^*(\Xi). \tag{8.8}$$

---

* COLEMAN [3] used jump continuations of general histories in his thermodynamics of materials with memory to show that simple materials with gradually fading memory behave as perfectly elastic materials (with stored energy) in sufficiently fast processes.

** To prove (8.7), one uses a continuity in mean theorem for translation operators over $\mathscr{L}_2$-space; we deliberately omit the details, which are standard.

We have just seen that for a given history F^t one can obtain the stress-strain curve $T^*(\cdot)$ for supposed impulses at time t by studying sequences of (continuous) linear continuations of F^t involving increasing rates of strain. Once one has $T^*(\Xi)$ for different values of Ξ, one can compute E_t and $\widetilde{E}_t$ using (8.4). It may be useful to the physicist to know that he can obtain the same results by differentiating directly the functions $T^{(\delta)}(\Xi)$ and extrapolating the result first to $\Xi=0$ and then to $\delta=0$; *i.e.*,

$$\lim_{\delta\to 0}\left[\lim_{\Xi\to 0}\frac{d}{d\,\Xi}\,T^{(\delta)}(\Xi)\right]=E_t, \tag{8.9a}$$

$$\lim_{\delta\to 0}\left[\lim_{\Xi\to 0}\frac{d^2}{d\,\Xi^2}\,T^{(\delta)}(\Xi)\right]=\widetilde{E}_t. \tag{8.9b}$$

To justify (8.9a), we let $f^{(\delta)}$ be that function on $[0,\infty)$ whose value for each s is the partial derivative of $F^{t+\delta,\Xi}(s)$ with respect to Ξ. Then, by (8.5), we have

$$f^{(\delta)}(s)=\begin{cases}\dfrac{\delta-s}{\delta}, & 0\leqq s<\delta,\\ 0, & \delta\leqq s<\infty.\end{cases} \tag{8.10}$$

An easy calculation, using (1.4) and (1.5), tells us that $f^{(\delta)}$ is in $\mathscr{H}_h^{\oplus}$, and, further,

$$\lim_{\zeta\to 0}\left\|f^{(\delta)}-\frac{F^{t+\delta,\Xi+\zeta}-F^{t+\delta,\Xi}}{\zeta}\right\|_h^{\oplus}=0; \tag{8.11}$$

i.e., $f^{(\delta)}$ is also the derivative with respect to Ξ of $F^{t+\delta,\Xi}$ regarded as a function of Ξ with values in $\mathscr{H}_h^{\oplus}$. It follows from (8.11), our smoothness hypotheses on $\mathfrak{T}$, and a familiar chain rule, that $\frac{d}{d\,\Xi}\,T^{(\delta)}$ exists and is given by

$$\frac{d}{d\,\Xi}\,T^{(\delta)}(\Xi)=D\mathfrak{T}(F^{t+\delta,\Xi})\,f^{(\delta)}(0)+\delta\mathfrak{T}(F^{t+\delta,\Xi}\,|\,f_r^{(\delta)}), \tag{8.12}$$

where $f_r^{(\delta)}$ is the restriction to $(0,\infty)$ of $f^{(\delta)}$. We now note that $f^{(\delta)}$ is independent of Ξ and

$$f^{(\delta)}(0)=1. \tag{8.13}$$

Furthermore, by (8.5),

$$\lim_{\Xi\to 0}\|F^{t+\delta,\Xi}-F^{t+\delta,0}\|_h^{\oplus}=0 \tag{8.14a}$$

where $F^{t+\delta,0}$ is the *static continuation* of F^t by amount δ:*

$$F^{t+\delta,0}=\begin{cases}F^t(0), & 0\leqq s<\delta,\\ F^t(s-\delta), & \delta\leqq s<\infty.\end{cases} \tag{8.14b}$$

---

* Static continuations of histories of deformation gradients were defined and studied by Coleman & Noll [*16*] in their work on the stress-relaxation theorem. Wang [*17*] has recently found more general statements of that theorem and has explored the behavior as $\delta\to\infty$ of the norms of static continuations in various function spaces. Coleman [*18*] has pointed out that if a static continuation is isothermal, then the Helmholtz free energy decreases monotonically to a limiting (equilibrium) value as $\delta\to\infty$.

Hence, since $D\mathfrak{T}(F^{t+\delta,\Xi})$ and $\delta\mathfrak{T}(F^{t+\delta,\Xi}|f_r^{(\delta)})$ are both continuous functions of $F^{t+\delta,\Xi}$, we can write

$$\lim_{\Xi\to 0}\frac{d}{d\Xi}T^{(\delta)}(\Xi)=D\mathfrak{T}(F^{t+\delta,0})+\delta\mathfrak{T}(F^{t+\delta,0}|f_r^{(\delta)}). \tag{8.15}$$

It is not difficult to show that

$$\lim_{\delta\to 0}\|f_r^{(\delta)}\|_h=0, \tag{8.16a}$$

and

$$\lim_{\delta\to 0}\|F^{t+\delta,0}-F^t\|_h^{\oplus}=0. \tag{8.16b}$$

Thus, since the first term on the right in (8.15) is continuous in $F^{t+\delta,0}$ and the second term is linear in $f_r^{(\delta)}$ and jointly continuous in $F^{t+\delta,0}$ and $f_r^{(\delta)}$, we have

$$\lim_{\delta\to 0}\left[\lim_{\Xi\to 0}\frac{d}{d\Xi}T^{(\delta)}(\Xi)\right]=D\mathfrak{T}(F^t), \tag{8.17}$$

which equation, by (1.8), is the same as (8.9a).

A similar argument justifies (8.9b): Since the second derivative of $F^{t+\delta,\Xi}$ with respect to Ξ is zero, differentiation of (8.12) with respect to Ξ yields

$$\begin{aligned}\frac{d^2}{d\Xi^2}T^{(\delta)}(\Xi)&=D^2\mathfrak{T}(F^{t+\delta,\Xi})\,f^{(\delta)}(0)^2+\\&+2\delta\,D\mathfrak{T}(F^{t+\delta,\Xi}|f^{(\delta)})\,f^{(\delta)}(0)+\delta^2\mathfrak{T}(F^{t+\delta,\Xi}|f_r^{(\delta)},f_r^{(\delta)}),\end{aligned} \tag{8.18}$$

and, by (8.13), (8.14), and the continuity of $D^2\mathfrak{T}$ and $\delta D\mathfrak{T}$,

$$\lim_{\Xi\to 0}\frac{d^2}{d\Xi^2}T^{(\delta)}(\Xi)=D^2\mathfrak{T}(F^{t+\delta,0})+2\delta\,D\mathfrak{T}(F^{t+\delta,0}|f_r^{(\delta)})+\delta^2\mathfrak{T}(F^{t+\delta,0}|f_r^{(\delta)},f_r^{(\delta)}). \tag{8.19}$$

Clearly, equation (8.16) and the continuity and linearity of the appropriate terms on the right in (8.19) give

$$\lim_{\delta\to 0}\left[\lim_{\Xi\to 0}\frac{d^2}{d\Xi^2}T^{(\delta)}(\Xi)\right]=D^2\mathfrak{T}(F^t), \tag{8.20}$$

which is, by (1.9), our desired result.

Of course, whenever F^t in (8.5) and (8.6) is the rest history $F_0^\dagger$ of (4.2), the moduli E_t and $\widetilde{E}_t$ obtained from (8.9) will be, respectively, the numbers $G(0)$ and $\widetilde{E}$ of (4.4b) and (4.4d).

9. $\widetilde{E}$ and Measurements of Wave Velocity

Consider a given homogeneously strained material with two different histories F^t and $F_0^\dagger$. Let $F_0^\dagger$ be the rest history of (4.2), and let F^t be close to $F_0^\dagger$ in the sense that the relative strain $\gamma^t(s)$ defined by $\gamma^t(s)=\gamma(t-s)=F^t(s)-F_0$ is small in magnitude for all s; i.e., put

$$\gamma^t=F^t-F_0^\dagger, \tag{9.1}$$

and assume that

$$\varepsilon=\sup_{0\leqq s<\infty}|\gamma^t(s)| \tag{9.2}$$

is small. If we let U and U^0 be the intrinsic velocities of waves propagating into the material when the histories of the region in front of the waves are, respectively, F^t and $F_0^\dagger$, by (2.14) and (2.15) we have

$$\varrho_R (U^0)^2 = D\mathfrak{T}(F_0^\dagger) = G(0) = E, \tag{9.3a}$$

and

$$\varrho_R U(t)^2 = D\mathfrak{T}(F^t). \tag{9.3b}$$

It follows from (9.1) and our smoothness hypotheses on $\mathfrak{T}$ that

$$D\mathfrak{T}(F^t) - D\mathfrak{T}(F_0^\dagger) = D^2\mathfrak{T}(F_0^\dagger)\,\gamma(t) + \delta D\mathfrak{T}(F_0^\dagger\,|\,\gamma_r^t) + o\left(\|\gamma^t\|_h^\oplus\right), \tag{9.4}$$

where γ_r^t is the restriction of γ^t to $(0, \infty)$, $\gamma(t) = \gamma^t(0)$, and $\delta D\mathfrak{T}$ and $D^2\mathfrak{T}$ are defined in (1.7). Since $\|\gamma^t\|_h^\oplus = O(\varepsilon)$, the first two terms on the right in (9.4) are $O(\varepsilon)$ and

$$o\left(\|\gamma^t\|_h^\oplus\right) = o(\varepsilon). \tag{9.5}$$

We note that the coefficient $D^2\mathfrak{T}(F_0^\dagger)$ of $\gamma(t)$ is the second-order modulus $\widetilde{E}$ of (4.4d). Furthermore, since $\delta D\mathfrak{T}(F_0^\dagger\,|\,\cdot)$ is a linear continuous functional over the Hilbert space $\mathscr{H}_h$, it can be represented as an inner-product with a function $\widetilde{K}$ in $\mathscr{H}_h$:

$$\delta D\mathfrak{T}(F_0^\dagger\,|\,\gamma_r^t) = \int_0^\infty \widetilde{K}(s)\,\gamma_r^t(s)\,h(s)^2\,ds. \tag{9.6}$$

Hence, letting the function $\widetilde{G}$ be the unique solution of

$$\frac{d}{ds}\widetilde{G}(s) = \widetilde{K}(s)\,h(s)^2, \tag{9.7a}$$

$$\widetilde{G}(0) = \widetilde{E} = D^2\mathfrak{T}(F_0^\dagger), \tag{9.7b}$$

we can write (9.4) in the form

$$D\mathfrak{T}(F^t) - D\mathfrak{T}(F_0^\dagger) = \widetilde{G}(0)\,\gamma(t) + \int_0^\infty \widetilde{G}'(s)\,\gamma_r^t(s)\,ds + o(\varepsilon). \tag{9.8}$$

It follows from (9.3) and (9.8) that if we put

$$\Delta = \varrho_R[U(t)^2 - (U^0)^2], \tag{9.9}$$

we can write

$$\Delta = \widetilde{G}(0)\,\gamma(t) + \int_0^\infty \widetilde{G}'(s)\,\gamma(t-s)\,ds + o(\varepsilon). \tag{9.10}$$

We have just seen that for an acceleration wave entering a material whose strain relative to the configuration $x = F_0 X + X_0$ has always been small, the quantity Δ of (9.9) is given, to a first approximation, by an equation of precisely the same form as the constitutive equation for the stress in linear viscoelasticity, with the material function $\widetilde{G}$, defined by (9.6) and (9.7), playing the role of the stress-relaxation function G. We call $\widetilde{G}$ the *wave-velocity relaxation function corresponding* to $F_0^\dagger$. Whenever measurement of stress can be replaced by measurement of Δ, an experimental program for determining $G(s)$ for small s can be converted to one for determining $\widetilde{G}(s)$ for small s and hence of $\widetilde{E} = \lim_{s\to 0} \widetilde{G}(s)$.

For an example we consider the analogue of a stress-relaxation experiment and suppose that γ has the special time-dependence

$$\gamma(t) = \begin{cases} 0, & t < 0, \\ \gamma^*, & t \geqq 0. \end{cases} \tag{9.11}$$

Then, for $t>0$ we have

$$\gamma^t(s)=\begin{cases}\gamma^*, & 0\leqq s\leqq t,\\ 0, & s>t,\end{cases} \tag{9.12}$$

and equation (9.10) yields

$$\Delta=\gamma^*\widetilde{G}(t)+o(\gamma^*). \tag{9.13}$$

If U^0 and $U(t)$ corresponding to (9.11) are known, then, by (9.7b) and (9.13),

$$\widetilde{E}=\lim_{t\to 0}\left[\lim_{\gamma^*\to 0}\frac{U(t)^2-(U^0)^2}{\varrho_R\,\gamma^*}\right]. \tag{9.14}$$

The support of this research by the National Science Foundation and the Advanced Research Projects Agency of the Department of Defense is gratefully acknowledged.

References

[1] Coleman, B. D., M. E. Gurtin, & I. Herrera R., Arch. Rational Mech. Anal. **19**, 1—18 (1965).
[2] Noll, W., Arch. Rational Mech. Anal. **2**, 197—226 (1958).
[3] Coleman, B. D., Arch. Rational Mech. Anal. **17**, 230—254 (1964).
[4] Truesdell, C., & R. A. Toupin, The Classical Field Theories. In the Encyclopedia of Physics, Vol. III/1, edited by S. Flügge. Berlin-Göttingen-Heidelberg: Springer 1960.
[5] Thomas, T. Y., Plastic Flow and Fracture in Solids. New York-London: Academic Press 1961.
[6] Chu, B.-T., Stress Waves in Isotropic Viscoelastic Materials 1, a report to ARPA from the Division of Engineering, Brown University, Providence, March 1962.
[7] Gurtin, M. E., & I. Herrera R., Quart. Appl. Math. **22**, 360—364 (1964).
[8] Thomas, T. Y., J. Math. Mech. **6**, 455—469 (1957).
[9] Green, W. A., Arch. Rational Mech. Anal. **16**, 79—89 (1964).
[10] Bland, D. R., J. Mech. Phys. Solids **12**, 245—267 (1964).
[11] Coleman, B. D., & W. Noll, Ann. N.Y. Acad. Sci. **89**, 672—714 (1961).
[12] Coleman, B. D., Arch. Rational Mech. Anal., forthcoming.
[13] Truesdell, C., Arch. Rational Mech. Anal. **8**, 263—296 (1961).
[14] Gross, B., Mathematical Structure of the Theories of Viscoelasticity. Paris: Hermann & Cie. 1953.
[15] Ferry, J. D., Viscoelastic Properties of Polymers. New York: Wiley 1961.
[16] Coleman, B. D., & W. Noll, Simple Fluids with Fading Memory. In: Proc. Internat. Sympos. Second-Order Effects, Haifa, 1962: New York: Macmillan 1964, pp. 530—552.
[17] Wang, C.-C., Arch. Rational Mech. Anal. **18**, 117—126 (1965).
[18] Coleman, B. D., Arch. Rational Mech. Anal. **17**, 1—46 (1964).
[19] Lee, E. H., & I. Kanter, J. Appl. Phys. **24**, 1115—1122 (1953).
[20] Hunter, S. C., Viscoelastic Waves. In: Progress in Solid Mechanics, Vol. I, edited by I. N. Sneddon & R. Hill. Amsterdam: North-Holland Publishing Co. 1960.

Mellon Institute
Pittsburgh, Pennsylvania
and
Brown University
Providence, Rhode Island

(Received January 23, 1965)

III. Thermodynamic Influences on the Growth and Decay of Acceleration Waves

BERNARD D. COLEMAN & MORTON E. GURTIN

Contents

Preface

Previous papers [*1*, *2*] in this series have dealt with shock and acceleration waves in materials with memory.* Those articles, resting on the constitutive assumption that the stress is a functional of the history of the deformation gradient, neglected thermal influences. In this and subsequent papers we shall allow the stress to be affected not only by the history of the strain, but also the history of a thermodynamic variable such as the temperature. We shall make no restriction on this influence of the history of strain and temperature on the stress other than to suppose that it be compatible with the laws of thermodynamics and the principle of fading memory.**

In the present article we calculate the velocity of one-dimensional acceleration waves and show how the dissipative effects of heat conduction and internal mechanical damping affect the amplitude of such waves. Papers I and II contain examples of circumstances in which the behavior of a three-dimensional body is described by a one-dimensional theory; those examples apply here with only trivial modification. We do detailed calculations for two types of materials: The first, called *definite conductors*, have the property that the derivative of

* We shall refer to references [*1*] and [*2*] as "I" and "II", respectively.

** In this and paper IV we shall make frequent use of the results in COLEMAN'S two papers [*3*], [*4*] on the thermodynamics of materials with memory.

the heat flux with respect to the temperature gradient, though it may depend on the history of the strain and temperature, is always strictly negative; the second type consists in *non-conductors*, for which the heat flux is always zero. Acceleration waves in definite conductors are always homothermal, while for non-conductors they are homentropic. For both types of materials one can calculate explicitly, without further specializing assumptions, the dependence on time of the amplitude of a wave propagating through a homogeneous medium previously at rest.

1. Basic Assumptions

As in most of I and II, the kinematical and mechanical concepts used here are one-dimensional. A motion is described by a function $x = x(X, t)$ giving the location x at time t of the material point which has the position X in a reference configuration $\mathscr{R}$ with density ϱ_R. It is supposed that not only the stress T, but also the absolute temperature ϑ (which is always > 0), the specific internal energy ε (in units of energy per unit mass), the specific entropy η, the heat flux q, the specific extrinsic body force b, and the specific extrinsic heat supply* r may all be regarded as functions of X and t. The law of balance of momentum takes the form**

$$\frac{d}{dt}\int_{X_\alpha}^{X_\beta} \dot{x}\, \varrho_R\, dX = \int_{X_\alpha}^{X_\beta} b\, \varrho_R\, dX + T(X_\beta, t) - T(X_\alpha, t), \tag{1.1}$$

and for the law of balance of energy we write

$$\begin{aligned}\frac{d}{dt}\int_{X_\alpha}^{X_\beta} \left(\frac{\dot{x}^2}{2} + \varepsilon\right) \varrho_R\, dX = \int_{X_\alpha}^{X_\beta} (\dot{x}\, b + r)\, \varrho_R\, dX + T(X_\beta, t)\, \dot{x}(X_\beta, t) - \\ - T(X_\alpha, t)\, \dot{x}(X_\alpha, t) - q(X_\beta, t) + q(X_\alpha, t);\end{aligned} \tag{1.2}$$

a superimposed dot denotes the material time derivative; i.e., $\dot{x} = \partial_t x(X, t)$ is the velocity. The equations (1.1) and (1.2) are assumed to hold at all times t and for every pair X_α, X_β of points in $\mathscr{R}$. Of course, the Helmholtz free energy ψ defined by

$$\psi = \varepsilon - \vartheta\eta \tag{1.3}$$

will also be a function of X and t, and so will the deformation gradient $F = \partial_X x(X, t)$.

The *histories* up to time t of the deformation gradient and the temperature at a fixed material point X are real-valued functions F^t, ϑ^t defined by

$$F^t(s) = F(t - s) = \partial_X x(X, t - s), \qquad 0 \leqq s < \infty, \tag{1.4a}$$

$$\vartheta^t(s) = \vartheta(t - s) = \vartheta(X, t - s), \qquad 0 \leqq s < \infty. \tag{1.4b}$$

To emphasize that these histories depend on X as well as t, we sometimes write $F^t(X, \cdot)$ for F^t and $\vartheta^t(X, \cdot)$ for ϑ^t. Note that the domain of definition of F^t and

---

* In most applications r is assumed to be zero.

** This is same as (3.8) of I except that the ϱ_0 shown there is written ϱ_R here and in II.

ϑ^t is the entire positive real axis $[0, \infty)$. The restrictions F_r^t and ϑ_r^t of these functions to the open interval $(0, \infty)$ are called *past histories*, while the numbers $F(t) = F^t(0)$ and $\vartheta(t) = \vartheta(0)$ are called *present values*.* We denote the present value of the temperature gradient at X by $g(t)$:

$$g(t) = \partial_X \vartheta(X, t). \tag{1.5}$$

In accordance with the thermodynamic theory of [3], we now lay down the following basic *constitutive assumptions*. We assume that the material under consideration is characterized by four functionals, $\mathfrak{p}, \mathfrak{T}, \mathfrak{h}, \mathfrak{q}$. The first three of these material functionals give the present values of ψ, T, and η at a material point whenever the histories F^t and ϑ^t are specified; the last functional gives the present value of q when, in addition to F^t and ϑ^t, $g(t)$ is specified:

$$\psi(t) = \mathfrak{p}(F^t, \vartheta^t), \tag{1.6a}$$

$$T(t) = \mathfrak{T}(F^t, \vartheta^t), \tag{1.6b}$$

$$\eta(t) = \mathfrak{h}(F^t, \vartheta^t), \tag{1.6c}$$

$$q(t) = \mathfrak{q}\big(F^t, \vartheta^t; g(t)\big). \tag{1.6d}$$

For convenience we assume that the material is *homogeneous* and that a homogeneous configuration $\mathscr{R}$ has been taken as reference; thus, although F^t, ϑ^t, g, ψ, T, η, and q generally depend on both X and t, the functionals $\mathfrak{p}, \mathfrak{T}, \mathfrak{h}, \mathfrak{q}$ in (1.6) and the mass density ϱ_R corresponding to $\mathscr{R}$ are independent of X.

We also assume that the functionals $\mathfrak{p}, \mathfrak{T}, \mathfrak{h}, \mathfrak{q}$ obey the principle of fading memory. To state this principle in our present context, we let h be a fixed *influence function*, i.e., a positive, monotone-decreasing, continuous function on $[0, \infty)$ decaying to zero fast enough to be square integrable, and we define the norm $\|\Lambda^t\|^{\oplus}$ of a pair $\Lambda^t = (F^t, \vartheta^t)$ of histories to be

$$\|\Lambda^t\|^{\oplus} = \|F_r^t\|_h + \|\vartheta_r^t\|_h + |F^t(0)| + |\vartheta^t(0)|; \tag{1.7}$$

here $|\cdot|$ denotes the absolute value and $\|\cdot\|_h$ the "h-norm":

$$\|f\|_h = \sqrt{\int_0^\infty |f(s)|^2 h(s)^2\, ds}. \tag{1.8}$$

We assume that there exists an influence function h such that $\mathfrak{p}, \mathfrak{T}$, and $\mathfrak{h}$, have for their common domain of definition an open subset $\mathscr{D}$ of the function space $\mathscr{S}$ composed of those function pairs $\Lambda^t = (F^t, \vartheta^t)$ whose norm (1.7) is finite.** We further assume that $\mathfrak{p}, \mathfrak{T}$, and $\mathfrak{h}$ are smooth functions over $\mathscr{D}$ with respect to the norm $\|\cdot\|^{\oplus}$. To be more specific, let $\mathfrak{f}$ be a functional representing either $\mathfrak{p}, \mathfrak{T}$, or $\mathfrak{h}$. We are assuming here that, for each Λ^t in $\mathscr{D}$, $\mathfrak{f}$ has n^{th} order Fréchet derivatives $d^n\mathfrak{f}(\Lambda^t|\cdot, \ldots, \cdot)$ for $n = 1, \ldots, N$; these derivatives are bounded, symmetric, n-linear forms over $\mathscr{S}$ with the property that for all functions Ω in $\mathscr{S}$

* *Cf.* § 9 of [4].

** The space $\mathscr{S}$ equals the Cartesian product $\mathscr{H}_h^{\oplus} \times \mathscr{H}_h^{\oplus}$ where $\mathscr{H}_h^{\oplus}$ is the function space discussed in § 1 of I. Actually, both $\mathscr{S}$ and $\mathscr{H}_h^{\oplus}$ have for their elements not functions but rather equivalence classes of functions differing only on sets of measure zero; such technicalities do not, however, interest us here.

with $\Lambda^t+\Omega$ in $\mathscr{D}$

$$\mathfrak{f}(\Lambda^t+\Omega)=\mathfrak{f}(\Lambda^t)+\sum_{n=1}^{N}\frac{1}{n!}\,d^n\mathfrak{f}(\Lambda^t|\Omega,\Omega,\ldots,\Omega)+o(\|\Omega\|^{\oplus N}). \tag{1.9}$$

Each functional $d^n\mathfrak{f}(\cdot|,\ldots,\cdot)$ is assumed to be jointly continuous in all of its $n+1$ arguments. For our present discussion we do not need Fréchet derivatives of all orders; it suffices to have $N=3$ when $\mathfrak{f}=\mathfrak{p}$, and $N=2$ when $\mathfrak{f}=\mathfrak{T}$ or $\mathfrak{h}$. We also assume that the functional $\mathfrak{q}(\cdot,\cdot,\cdot)$ of (1.6d) is continuously differentiable, in the sense of Fréchet, on $\mathscr{D}\times(-\infty,\infty)$.

We shall study in detail acceleration waves in two general types of materials: definite conductors of heat and non-conductors. These are defined as follows. A material is a *definite conductor* if the *heat conduction modulus* k_t, defined by

$$k_t=-\frac{\partial}{\partial g}\,\mathfrak{q}(F^t,\vartheta^t;g), \tag{1.10}$$

is always (strictly) positive:

$$k_t>0. \tag{1.11}$$

A material is a *non-conductor* if $\mathfrak{q}$ is the constant functional with value zero, *i.e.*, if

$$q\equiv 0. \tag{1.12}$$

In our treatment of definite conductors we shall *not* assume that q is a linear function of g, i.e., that

$$\mathfrak{q}(F^t,\vartheta^t,g)=-\mathfrak{K}(F^t,\vartheta^t)\,g. \tag{1.13}$$

We note, however, that a necessary and sufficient condition for a "Fourier-type material" (1.13) to be a definite conductor is that $\mathfrak{K}(F^t,\vartheta^t)$ be >0 for all F^t,ϑ^t. Of course, such a material is a non-conductor when $\mathfrak{K}(F^t,\vartheta^t)\equiv 0$.

2. Consequences of Fading Memory

Since a knowledge of the past histories F_r^t, ϑ_r^t, and the present values $F(t)$, $\vartheta(t)$ is equivalent to a knowledge of the pair $(F^t,\vartheta^t)=\Lambda^t$, we can write the constitutive equations (1.6) in the forms

$$\psi=\mathfrak{p}(F_r^t,\vartheta_r^t;F,\vartheta), \tag{2.1 a}$$

$$T=\mathfrak{T}(F_r^t,\vartheta_r^t;F,\vartheta), \tag{2.1 b}$$

$$\eta=\mathfrak{h}(F_r^t,\vartheta_r^t;F,\vartheta), \tag{2.1 c}$$

$$q=\mathfrak{q}(F_r^t,\vartheta_r^t;F,\vartheta,g); \tag{2.1 d}$$

here, for short, we have put F for $F(t)$, ϑ for $\vartheta(t)$, ψ for $\psi(t)$, *etc.* Letting $\mathfrak{f}$ stand for $\mathfrak{p}$, $\mathfrak{T}$, $\mathfrak{h}$, or $\mathfrak{q}$ at a fixed value of g, we write

$$\mathfrak{f}(F^t,\vartheta^t)=\mathfrak{f}(F_r^t,\vartheta_r^t;F,\vartheta). \tag{2.2}^\star$$

Among the implications of the principle of fading memory, as outlined in the preceding paragraph, is the existence of linear differential operators** D_F, D_ϑ, δ_F, δ_ϑ, which when they operate on $\mathfrak{f}$ give functionals $D_F\mathfrak{f}$, $D_\vartheta\mathfrak{f}$, $\delta_F\mathfrak{f}$, $\delta_\vartheta\mathfrak{f}$, defined

* A treatise on functional analysis would most likely distinguish, by the use of separate symbols, the two *different* functionals which we denote by $\mathfrak{f}$ in (2.2).

** The significance of these operators in thermodynamics was pointed out in [3], § 7; we here use the simpler definitions of [4], § 9. See also § 2 of I and § 1 of II.

by the following formulae:

$$D_F \mathfrak{f}(F^t, \vartheta^t) \stackrel{\text{def}}{=} \frac{\partial}{\partial F} \mathfrak{f}(F^t_r, \vartheta^t_r; F, \vartheta), \tag{2.3 a}$$

$$D_\vartheta \mathfrak{f}(F^t, \vartheta^t) \stackrel{\text{def}}{=} \frac{\partial}{\partial \vartheta} \mathfrak{f}(F^t_r, \vartheta^t_r; F, \vartheta), \tag{2.3 b}$$

$$\delta_F \mathfrak{f}(F^t, \vartheta^t | f) \stackrel{\text{def}}{=} \left[\frac{\partial}{\partial \nu} \mathfrak{f}(F^t_r + \nu f, \vartheta^t_r; F, \vartheta)\right]_{\nu=0}, \tag{2.4 a}$$

$$\delta_\vartheta \mathfrak{f}(F^t, \vartheta^t | f) \stackrel{\text{def}}{=} \left[\frac{\partial}{\partial \nu} \mathfrak{f}(F^t_r, \vartheta^t_r + \nu f; F, \vartheta)\right]_{\nu=0}. \tag{2.4 b}$$

These equations hold for all pairs F^t, ϑ^t in $\mathscr{D}$ and all functions f in the Hilbert space $\mathscr{H}_h$ of functions with finite h-norm (1.8). Each of the functionals on the left in (2.4) is jointly continuous in all of its arguments. Furthermore, for each pair F^t, ϑ^t, the functionals $\delta_F \mathfrak{f}$ and $\delta_\vartheta \mathfrak{f}$, as functions of f, are not only continuous with respect to the norm $\|\cdot\|_h$, but are also *linear* in f.

Our fading memory assumption implies also the existence and continuity of higher derivatives of $\mathfrak{f}$ such as $D_F^2 \mathfrak{f}$, $D_\vartheta D_F \mathfrak{f} = D_F D_\vartheta \mathfrak{f}$, $D_\vartheta^2 \mathfrak{f}$, $\delta_F D_F \mathfrak{f} = D_F \delta_F \mathfrak{f}$, $\delta_F D_\vartheta \mathfrak{f} = D_\vartheta \delta_F \mathfrak{f}$, $\delta_\vartheta D_F \mathfrak{f} = D_F \delta_\vartheta \mathfrak{f}$, $\delta_\vartheta D_\vartheta \mathfrak{f} = D_\vartheta \delta_\vartheta \mathfrak{f}$, $\delta_F^2 \mathfrak{f}$, and $\delta_\vartheta^2 \mathfrak{f}$. The following examples of formulae characterizing products of the differential operators D_F, D_ϑ, δ_F, δ_ϑ make it apparent that any two of these operators commute:

$$\begin{aligned} D_\vartheta D_F \mathfrak{f}(F^t, \vartheta^t) &\stackrel{\text{def}}{=} \frac{\partial^2}{\partial \vartheta\, \partial F} \mathfrak{f}(F^t_r, \vartheta^t_r; F, \vartheta) \\ &= \frac{\partial^2}{\partial F\, \partial \vartheta} \mathfrak{f}(F^t_r, \vartheta^t_r; F, \vartheta) \stackrel{\text{def}}{=} D_F D_\vartheta \mathfrak{f}(F^t, \vartheta^t), \end{aligned} \tag{2.5 a}$$

$$\begin{aligned} \delta_F D_F \mathfrak{f}(F^t, \vartheta^t | f) &\stackrel{\text{def}}{=} \left[\frac{\partial^2}{\partial \nu\, \partial F} \mathfrak{f}(F^t_r + \nu f, \vartheta^t_r; F, \vartheta)\right]_{\nu=0} \\ &= \frac{\partial}{\partial F}\left[\frac{\partial}{\partial \nu} \mathfrak{f}(F^t_r + \nu f, \vartheta^t_r; F, \vartheta)\right]_{\nu=0} \stackrel{\text{def}}{=} D_F \delta_F \mathfrak{f}(F^t, \vartheta^t | f), \end{aligned} \tag{2.5 b}$$

$$\begin{aligned} \delta_\vartheta \delta_F \mathfrak{f}(F^t, \vartheta^t | f, l) &\stackrel{\text{def}}{=} \left[\frac{\partial}{\partial \mu}\left[\frac{\partial}{\partial \nu} \mathfrak{f}(F^t_r + \nu f, \vartheta^t_r + \mu l; F, \vartheta)\right]_{\nu=0}\right]_{\mu=0} \\ &= \left[\frac{\partial}{\partial \nu}\left[\frac{\partial}{\partial \mu} \mathfrak{f}(F^t_r + \nu f, \vartheta^t_r + \mu l; F, \vartheta)\right]_{\mu=0}\right]_{\nu=0} \stackrel{\text{def}}{=} \delta_F \delta_\vartheta \mathfrak{f}(F^t, \vartheta^t | l, f); \end{aligned} \tag{2.5 c}$$

here f and l are arbitrary functions over $(0, \infty)$ with finite h-norm. It is easily shown that for each pair F^t, ϑ^t the quantities $\delta_F^2 \mathfrak{f}(F^t, \vartheta^t | f, g)$ and $\delta_\vartheta^2 \mathfrak{f}(F^t, \vartheta^t | f, l)$, are symmetric bilinear functions of f and l.

The derivatives $D_F \mathfrak{f}$ and $D_\vartheta \mathfrak{f}$ defined in (2.3) are "instantaneous-moduli," *i.e.*, partial derivatives with respect to present values, computed holding the past history fixed.* Certain of these moduli occur frequently in our theory. The number

$$E_i = D_F \mathfrak{T}(F^t, \vartheta^t) \tag{2.6}$$

may be called the *instantaneous stress-strain modulus at fixed present temperature* (corresponding to the histories F^t and ϑ^t); it equals the initial slope of the stress-strain curve for instantaneous response to isothermal strain impulses super-

* For a more detailed discussion see § 5 of [4].

imposed on F^t, ϑ^t at time t.* The number

$$\widetilde{E}_t = D_F^2 \mathfrak{T}(F^t, \vartheta^t), \tag{2.7}$$

which we call the *instantaneous second-order stress-strain modulus at fixed present temperature,* measures the initial curvature of the isothermal stress-strain curve for instantaneous response. We call

$$\alpha_t = D_\vartheta \mathfrak{T}(F^t, \vartheta^t) \tag{2.8}$$

the *instantaneous stress-temperature modulus at fixed present strain.* Also of importance here is the *instantaneous heat-strain modulus* at *fixed ϑ and g,* defined as

$$\varphi_t = D_F \mathfrak{q}(F^t, \vartheta^t; g). \tag{2.9}**$$

We note that for a general nonlinear material the instantaneous moduli E_t, $\widetilde{E}_t$ and α_t depend on the histories F^t and ϑ^t. Furthermore, φ_t and the ordinary heat conduction modulus k_t depend on not only F^t and ϑ^t but also on $g(t)$. Hence in a thermodynamic process $E_t, \widetilde{E}_t, \alpha_t, \varphi_t$, and k_t may change in time at a given material point; the subscript t reminds us of this possibility. When we wish to emphasize the dependence of these moduli on X at a fixed time we write $E_X, \widetilde{E}_X, \ldots$.

Since $\delta_F \mathfrak{T}(F^t, \vartheta^t | \cdot)$ is, for each pair F^t, ϑ^t, a linear continuous functional over the Hilbert space $\mathscr{H}_h$ with norm $\|\cdot\|_h$, this functional can be represented as an inner-product with a function K_t in $\mathscr{H}_h$:

$$\delta_F \mathfrak{T}(F^t, \vartheta^t | f) = \int_0^\infty K_t(s)\, f(s)\, h(s)^2\, ds. \tag{2.10}$$

Of course, K_t depends on F^t and ϑ^t; putting

$$G_t'(s) = K_t(s)\, h(s)^2, \tag{2.11}$$

we can make this dependence explicit by writing

$$G_t'(s) = \mathfrak{G}'(F^t, \vartheta^t; s). \tag{2.12}$$

We assume here that $\mathfrak{G}'(\cdot, \cdot; s)$ is, for each fixed s, a continuous functional over $\mathscr{D}$ and that, for each fixed pair F^t, ϑ^t in $\mathscr{D}$, $G_t'(s) = \mathfrak{G}'(F^t, \vartheta^t; s)$ is a differentiable function of s for $0 \leqq s < \infty$. Putting $G_t''(s) = \frac{d}{ds} G_t'(s)$, we further assume that the mapping $(F^t, \vartheta^t) \to G_t(\cdot)\, h(\cdot)^{-2}$ carries $\mathscr{D}$ into $\mathscr{H}_h$ and is continuous.

---

* In I and II, where the only material functional was $\mathfrak{T}$ and the temperature dependence of T was not considered at all, we called E_t the "instantaneous tangent modulus corresponding to F^t".

** In applications we expect to have $\mathfrak{q}(F^t, \vartheta^t, 0) = 0$, and hence $\varphi_t = 0$ whenever $g = 0$. Clearly, however, when $g \neq 0$ we can have $\varphi_t \neq 0$, even when (1.6d) has the special form (1.13).

In theories of *elastic* materials, it is customary to consider that special case of (1.13) in which $\mathfrak{K}(F^t, \vartheta^t)$ is just a function $K(F(t), \vartheta(t))$, of the present values $F(t)$ $\vartheta(t)$; then (2.9) reduces to $\varphi_t = -g\, \partial_F K(F(t), \vartheta(t))$ which, for $g < 0$, is positive or negative depending on whether the heat-conduction coefficient K increases or decreases with changing strain.

Let us consider the unique solution G_t of

$$\frac{d}{ds} G_t(s) = G_t'(s), \tag{2.13}$$

$$G_t(0) = E_t = D_F \mathfrak{T}(F^t, \vartheta^t). \tag{2.14}$$

This integral G_t of G_t' also depends on F^t, ϑ^t:

$$G_t(s) = \mathfrak{G}(F^t, \vartheta^t; s). \tag{2.15}$$

We call G_t the *stress-strain relaxation function* corresponding to F^t, ϑ^t.* The following remark, an obvious generalization of Remark 1.1 of II, motivates this nomenclature.

Remark 2.1. Let $F(\tau)$, $F^*(\tau)$, and $\vartheta(\tau)$, with $-\infty < \tau < \infty$, be such that for each t the pairs F^t, ϑ^t and F^{*t}, ϑ^t are in the domain $\mathscr{D}$ of $\mathfrak{T}$. Suppose that the *relative strain,*

$$\gamma(\tau) = F^*(\tau) - F(\tau), \tag{2.16}$$

is always small. Then, if we put

$$T(t) = \mathfrak{T}(F^t, \vartheta^t), \qquad T^*(t) = \mathfrak{T}(F^{*t}, \vartheta^t), \tag{2.17}$$

and

$$\epsilon = \sup_{-\infty<\tau<\infty} |\gamma(\tau)|, \tag{2.18}$$

we have, for all t,

$$T^*(t) = T(t) + G_t(0)\,\gamma(t) + \int_0^\infty G_t'(s)\,\gamma(t-s)\,ds + o(\epsilon), \tag{2.19}$$

where G_t is the stress-strain relaxation function corresponding to F^t, ϑ^t.

3. Thermodynamic Restrictions on the Constitutive Equations

By an *admissible thermodynamic process* we mean a specification of the fields $x, \vartheta, T, \eta, \varepsilon, q, b, r$, as functions of X and t, compatible with the balance laws (1.1), (1.2) and the constitutive equations (1.6). In our present subject the second law of thermodynamics reduces to the assertion that the Clausius-Duhem inequality,

$$\frac{d}{dt}\int_{X_\alpha}^{X_\beta} \eta\,\varrho_R\,dX \geqq \int_{X_\alpha}^{X_\beta} \frac{r}{\vartheta}\,\varrho_R\,dX + \frac{q(X_\alpha, t)}{\vartheta(X_\alpha, t)} - \frac{q(X_\beta, t)}{\vartheta(X_\beta, t)}, \tag{3.1}$$

must hold at all times t and for every pair X_α, X_β of points in $\mathscr{R}$. Working in a three-dimensional formulation and using the principle of fading memory in a form slightly weaker than that which we assume here**, COLEMAN*** has shown that in order for (3.1) to hold for all smooth admissible thermodynamic processes it is necessary and sufficient that the assertions in each of the following three remarks be true.

* To be more consistent with the terminology used in naming the instantaneous moduli E_t, $\widetilde{E}_t$, and α_t, we might call G_t the *stress-strain relaxation function at fixed temperature history* corresponding to F^t, ϑ^t. This appellation is, however, rather long.

** See our paragraph containing eqs. (1.6)–(1.9).

*** [3], Theorem 1.

Remark 3.1. The functionals $\mathfrak{p}$, $\mathfrak{T}$, and $\mathfrak{h}$ of (1.6) are not independent: $\mathfrak{p}$ determines $\mathfrak{T}$ through the *stress relation**,

$$\mathfrak{T}=\varrho_R D_F \mathfrak{p}, \tag{3.2}$$

and $\mathfrak{h}$ through the *entropy relation***

$$\mathfrak{h}=-D_\vartheta \mathfrak{p}. \tag{3.3}$$

Remark 3.2. The functional $\mathfrak{p}$ is such that the *internal dissipation* σ, which may be defined by either of the formulae***

$$\sigma(t)=\frac{1}{\vartheta(t)}\left[D_F\mathfrak{p}(F^t,\vartheta^t)\dot{F}+D_\vartheta\mathfrak{p}(F^t,\vartheta^t)\dot{\vartheta}-\frac{d}{dt}\mathfrak{p}(F^t,\vartheta^t)\right], \tag{3.4}$$

$$\sigma=\frac{1}{\vartheta}\left[\frac{1}{\varrho_R}T\dot{F}-\eta\dot{\vartheta}-\dot{\psi}\right], \tag{3.5}$$

obeys the following *internal dissipation inequality:*†

$$\sigma\geqq 0. \tag{3.6}$$

Whenever, $\frac{d}{ds}F^t(s)$, $\frac{d}{ds}\vartheta^t(s)$ exist (in a weak sense) for all $s\geqq 0$ and the functions $\frac{d}{ds}F^t_r$ and $\frac{d}{ds}\vartheta^t_r$ have finite h-norm, (3.4) may be written in the form

$$\sigma(t)=\sigma_F(t)+\sigma_\vartheta(t), \tag{3.7a}$$

$$\sigma_F(t)=\frac{1}{\vartheta(t)}\,\delta_F\mathfrak{p}\left(F^t,\vartheta^t\middle|\frac{d}{ds}F^t_r\right), \tag{3.7b}$$

$$\sigma_\vartheta(t)=\frac{1}{\vartheta(t)}\,\delta_\vartheta\mathfrak{p}\left(F^t,\vartheta^t\middle|\frac{d}{ds}\vartheta^t_r\right), \tag{3.7c}$$

which is equivalent to equation (7.11) of [3].††

Remark 3.3. The functional $\mathfrak{q}$ obeys the *general dissipation inequality*†††

$$\mathfrak{q}(F^t,\vartheta^t,g)\,g\leqq\varrho\,\vartheta^2\sigma \tag{3.8}$$

for all numbers g and all histories F^t, ϑ^t sufficiently tame that σ exists.

We assume that our materials obey the second law and hence that Remarks 3.1—3.3 hold.

We may note that since $\vartheta>0$ and the internal dissipation inequality tells us that σ is $\geqq 0$, in order for $\mathfrak{q}$ to obey the general dissipation inequality (3.8), it suffices that

$$\mathfrak{q}(F^t,\vartheta^t,g)\,g\leqq 0 \tag{3.9}$$

for all F^t, ϑ^t and g. This condition is trivially obeyed by a non-conductor. It follows from (1.10) and (1.11) that a definite conductor obeys (3.9) if, and only if,

$$\mathfrak{q}(F^t,\vartheta^t,0)=0 \tag{3.10}$$

* [3], eq. (7.13a).
** [3], eq. (7.13b).
*** [3], eq. (1.15); [2], eq. (4.3).
† [3], eq. (6.30).
†† See also eq. (9.8) of [4].
††† [3], eq. (6.25).

for all F^t, ϑ^t. Hence, if, in addition to (3.6) we also assume (3.10),* then we can assert that both non-conductors and definite conductors obey the general dissipation inequality.

Since, as we saw in (2.5a), the operators D_ϑ and D_F commute, we can write

$$D_\vartheta D_F \mathfrak{p} = D_F D_\vartheta \mathfrak{p}. \tag{3.11}$$

Substituting the stress relation (3.2) and the entropy relation (3.3) into (3.11), we obtain the important equation

$$D_\vartheta \mathfrak{T} = -\varrho_R D_F \mathfrak{h}, \tag{3.12}$$

which generalizes to materials with memory the equilibrium relation $\frac{\partial}{\partial\vartheta} T(F,\vartheta) = -\varrho_R \frac{\partial}{\partial F} \eta(F,\vartheta)$ familiar in the thermodynamic theory of rubber elasticity.**

Equivalent to (3.12) is

Remark 3.4. The instantaneous stress-temperature modulus defined in (2.8) is given by the equation

$$\alpha_t = -\varrho_R D_F \mathfrak{h}(F^t, \vartheta^t). \tag{3.13}$$

4. General Properties of Acceleration Waves

The *material representation of a wave**** is a smooth one-parameter family of points $Y_t \in \mathscr{R}$, $-\infty < t < \infty$, such that Y_t gives the material point (labeled by its position in $\mathscr{R}$) at which the wave† is to be found at time t. The *material trajectory* Σ of the wave is the set of ordered pairs Y_t, t with $-\infty < t < \infty$.

Here we call Y_t an *acceleration wave* if the fields $x(X,t)$ and $\vartheta(X,t)$ have the following properties.

A-1: x, $\dot{x}$, F, and ϑ are continuous functions of X and t jointly for all X and t, while $\ddot{x}$, $\dot{F}$, $\partial_X F$, $\dot{\vartheta}$, $\partial_X \vartheta$ and also $\dddot{x}$, $\ddot{F}$, $\partial_X \dot{F}$, $\partial_X^2 F$, $\ddot{\vartheta}$, $\partial_X \dot{\vartheta}$, $\partial_X^2 \vartheta$ have (at most) jump discontinuities across Σ but are continuous in X and t jointly everywhere else.

A-2††: $F_r^t(X,\cdot)$ and $\vartheta_r^t(X,\cdot)$ are smooth functions of X and t with respect to the norm $\|\cdot\|_h$ for all X and t.†††

We, of course, assume that for each X and t the pair $F^t(X,\cdot)$, $\vartheta^t(X,\cdot)$ is interior to the domain $\mathscr{D}$ of our constitutive functionals. We further assume that the extrinsic body force and heat supply are assigned in such a way that $b(X,t)$, $r(X,t)$, $\dot{b}(X,t)$, and $\partial_X r(X,t)$ are continuous functions of X for all t.

If $f(X,t)$, as a function of X, has only a jump discontinuity at $X = Y_t$, then we define the jump in $f(X,t)$ across Σ at time t by

$$[f](t) = f(Y_t^-, t) - f(Y_t^+, t). \tag{4.1}$$

---

* The conditions under which (3.10) holds in a general material with memory are discussed on p. 41 of [3].

** For example, eq. (12) on p. 441 of FLORY's treatise [7] is, at bottom, the equation $\frac{\partial}{\partial\vartheta} T(F,\vartheta) = -\varrho_R \frac{\partial}{\partial F} \eta(F,\vartheta)$ applied to strips at equilibrium with a tension T superimposed on a fixed ambient pressure.

*** *Cf.* I, § 3.

† Often called "the wave front".

†† In Remark 7.1 we observe that for a wave entering a homogeneous strict conductor at rest at constant temperature the assumption of A-1 *implies* A-2.

††† This assumption, explained in detail in § 2 of I, means that the mappings $X, t \to F_r^t(X,\cdot)$; $X, t \to \vartheta_r^t(X,\cdot)$ have values in $\mathscr{H}_h$ and are continuously differentiable when the norm $\|\cdot\|_h$ is used in $\mathscr{H}_h$.

It follows immediately from a famous theorem of MAXWELL* that A-1 implies the *compatibility conditions*

$$[\ddot{x}] = -U[\dot{F}] = U^2[\partial_X F], \tag{4.2}$$

and

$$[\dot{\vartheta}] = -U[\partial_X \vartheta], \tag{4.3}$$

with $U = \frac{d}{dt} Y_t$ the *intrinsic velocity* of the wave.

When our constitutive equations (1.6) and the smoothness given the constitutive functionals by the principle of fading memory are applied, it is found that A-1 and A-2 imply certain degrees of smoothness for ψ, T, η, ε, and q as functions of X and t. For example, it follows from A-1 and A-2 that the pair $F^t(X,\cdot)$, $\vartheta^t(X,\cdot)$ is, for all X and t, a continuous function of X and t with respect to the norm $\|\cdot\|^{\oplus}$ of (1.7). Hence, by (1.3), and the continuity in $\mathscr{S}$ of $\mathfrak{p}$, $\mathfrak{T}$, and $\mathfrak{h}$, the fields ψ, T, η, and ε must be continuous for all X and t:

$$[\psi] = [T] = [\eta] = [\varepsilon] = 0. \tag{4.4}$$

A similar argument shows that the derivatives $\dot{T}$, $\partial_X T$, $\dot{\varepsilon}$, $\partial_X \varepsilon$, $\dot{q}$, $\partial_X q$ are continuous functions of X and t whenever $X \neq Y_t$. Using such regularity conditions, one can easily justify**

Remark 4.1. For processes involving acceleration waves, the laws of balance of momentum and energy, (1.1) and (1.2), are together equivalent to the assertion that for $X \neq Y_t$

$$\partial_X T + \varrho_R b = \varrho_R \ddot{x}, \tag{4.5}$$

$$\varrho_R \dot{\varepsilon} = T\dot{F} - \partial_X q + \varrho_R r, \tag{4.6}$$

while for $X = Y_t$

$$[\partial_X T] = \varrho_R [\ddot{x}], \tag{4.7}$$

$$\varrho_R [\dot{\varepsilon}] = T[\dot{F}] - [\partial_X q], \tag{4.8}$$

$$[q] = 0. \tag{4.9}$$

On looking at (2.1a), we see that A-1, A-2, the smoothness properties of $\mathfrak{p}$, and a chain rule, imply that for $X \neq Y_t$ we can write

$$\dot{\psi} = D_F \mathfrak{p}(F^t, \vartheta^t)\, \dot{F} + D_\vartheta \mathfrak{p}(F^t, \vartheta^t)\, \dot{\vartheta} + \delta_F \mathfrak{p}(F^t, \vartheta^t | \dot{F}_r^t) + \delta_\vartheta \mathfrak{p}(F^t, \vartheta^t | \dot{\vartheta}_r^t). \tag{4.10}$$

Since the function pair F^t, ϑ^t, as a point in $\mathscr{S}$, and the functions $\dot{F}_r^t$ and $\dot{\vartheta}_r^t$, as points in $\mathscr{H}_h$, all depend continuously on X and t for every X and t, the jumps in the numbers $D_F \mathfrak{p}(F^t, \vartheta^t)$, $D_\vartheta \mathfrak{p}(F^t, \vartheta^t)$, $\delta_F \mathfrak{p}(F^t, \vartheta^t | \dot{F}_r^t)$, and $\delta_\vartheta \mathfrak{p}(F^t, \vartheta^t | \dot{\vartheta}_r^t)$ across Σ are zero, and (4.10) yields

$$[\dot{\psi}] = D_F \mathfrak{p}(F^t, \vartheta^t)\, [\dot{F}] + D_\vartheta \mathfrak{p}(F^t, \vartheta^t)\, [\dot{\vartheta}]. \tag{4.11}$$

---

* For a statement of MAXWELL's Theorem see § 175 of [5] or § 3 of I. Detailed derivations of (4.2) are given in § 190 of [5] and the sentence containing eqs. (5.1) and (5.2) of I.

** The arguments used are similar to those explained in § 5 of I and § 241 of [5]. The eq. (4.9) was put forth by FOURIER [6] in a special context.

Inserting the stress relation (3.2) and the entropy equation (3.3) into (4.11), we obtain the important equation:

$$[\dot{\psi}] = \frac{1}{\varrho_R} T[\dot{F}] - \eta [\dot{\vartheta}]. \tag{4.12}$$

The equations (1.3) and (4.4) imply

$$[\dot{\psi}] = [\dot{\varepsilon}] - \eta [\dot{\vartheta}] - \vartheta [\dot{\eta}], \tag{4.13}$$

and on subtracting this from (4.12) we get another useful equation:

$$[\dot{\varepsilon}] = \frac{1}{\varrho_R} T[\dot{F}] + \vartheta [\dot{\eta}]. \tag{4.14}$$

An acceleration wave with

$$[\partial_X \vartheta] = [\dot{\vartheta}] = 0 \tag{4.15}$$

is called *homothermal*; one with

$$[\partial_X \eta] = [\dot{\eta}] = 0 \tag{4.16}$$

is called *homentropic*.

From (4.12) and (4.14) we read off

Theorem 4.1. *For a homothermal acceleration wave*

$$T[\dot{F}] = \varrho_R [\dot{\psi}]; \tag{4.17}$$

for a homentropic acceleration wave

$$T[\dot{F}] = \varrho_R [\dot{\varepsilon}]. \tag{4.18}$$

Since the pair F^t, ϑ^t in $\mathscr{S}$ depends continuously on X and t, the jump in q across Σ is given by

$$[q] = \mathfrak{q}\left(F^t(Y_t, \cdot), \vartheta^t(Y_t, \cdot); g(Y_t^-, t)\right) - \mathfrak{q}\left(F^t(Y_t, \cdot), \vartheta^t(Y_t, \cdot); g(Y_t^+, t)\right). \tag{4.19}$$

For a definite heat conductor $\mathfrak{q}(F^t, \vartheta^t; g)$ is a strictly monotone function of g for each fixed pair F^t, ϑ^t. Hence, for such a material (4.19) tells us that whenever $[g] \neq 0$ we must have $[q] \neq 0$, but, since, by (4.9), $[q] \neq 0$ is impossible, we conclude that $[g] = 0$. It then follows from (4.3) that $[\dot{\vartheta}] = 0$. This proves*

Theorem 4.2. *Every acceleration wave in a definite conductor is homothermal.*

Since for general acceleration waves both $[\eta] = 0$ and $[q] = 0$, the argument giving (4.3) also gives

$$[\dot{\eta}] = -U[\partial_X \eta], \tag{4.20}$$

$$[\dot{q}] = -U[\partial_X q]. \tag{4.21}$$

Combination of (4.14) and (4.8) yields the relation

$$\varrho_R \vartheta [\dot{\eta}] = -[\partial_X q]. \tag{4.22}$$

Since $\varrho_R \vartheta > 0$, the equations (4.9), (4.22), and (4.21) yield

Remark 4.1. For every homentropic wave, in addition to $[q] = 0$ we have $[\dot{q}] = [\partial_X q] = 0$.

---

* The argument we have just given is a generalization of one given by TRUESDELL [8] for elastic materials obeying Fourier's law and attributed by TRUESDELL to FOURIER [6]; see also [5], § 296.

More important than this remark, however, is*

Theorem 4.3. *Every acceleration wave with $U \neq 0$ in a non-conductor is homentropic.*

Proof. In a non-conductor $q \equiv 0$; hence $[\partial_X q] = 0$, and, since $\varrho_R \vartheta > 0$, we infer from (4.22) that $[\dot{\eta}] = 0$. But, when $U \neq 0$ (4.20) tells us that $[\dot{\eta}] = 0$ implies $[\partial_X \eta] = 0$; q.e.d.

5. The Velocity and Amplitude of Homothermal Waves

It follows from A-1, A-2, the smoothness of $\mathfrak{T}$, and a chain rule that for $X \neq Y_t$ we can write

$$\begin{aligned}\partial_X T = D_F \mathfrak{T}(F^t, \vartheta^t)\, \partial_X F + D_\vartheta \mathfrak{T}(F^t, \vartheta^t)\, \partial_X \vartheta + \\ + \delta_F \mathfrak{T}(F^t, \vartheta^t \mid \partial_X F_r^t) + \delta_\vartheta \mathfrak{T}(F^t, \vartheta^t \mid \partial_X \vartheta_r^t).\end{aligned} \tag{5.1}$$

Since the pair F^t, ϑ^t and the functions $\partial_X F_r^t$ and $\partial_X \vartheta_r^t$, regarded as points in their appropriate function spaces, depend continuously on X and t for all X and t, the numbers $D_F \mathfrak{T}(F^t, \vartheta^t)$, $D_\vartheta \mathfrak{T}(F^t, \vartheta^t)$, $\delta_F \mathfrak{T}(F^t, \vartheta^t \mid \partial_X F_r^t)$, and $\delta_\vartheta \mathfrak{T}(F^t, \vartheta^t \mid \partial_X \vartheta_r^t)$ are continuous functions of X and t even at $X = Y_t$. Hence, (5.1) yields

$$[\partial_X T] = D_F \mathfrak{T}(F^t, \vartheta^t)\, [\partial_X F] + D_\vartheta \mathfrak{T}(F^t, \vartheta^t)\, [\partial_X \vartheta] \tag{5.2}$$

where the coefficients of $[\partial_X F]$ and $[\partial_X \vartheta]$ are to be evaluated at the point $X = Y_t$. Putting, in accordance with (2.6) and (2.8),

$$E_{Y_t} = D_F \mathfrak{T}\big(F^t(Y_t, \cdot), \vartheta^t(Y_t, \cdot)\big), \tag{5.3 a}$$

$$\alpha_{Y_t} = D_\vartheta \mathfrak{T}\big(F^t(Y_t, \cdot), \vartheta^t(Y_t, \cdot)\big), \tag{5.3 b}$$

we can write (5.2) in the form

$$[\partial_X T] = E_{Y_t} [\partial_X F] + \alpha_{Y_t} [\partial_X \vartheta]. \tag{5.4}$$

For a homothermal wave $[\partial_X \vartheta] = 0$, and (5.4) becomes

$$[\partial_X T] = E_{Y_t} [\partial_X F]. \tag{5.5}$$

Combining this equation with (4.7) and (4.2), we find that

$$(\varrho_R U_t^2 - E_{Y_t})\, [\partial_X F] = 0, \tag{5.6}$$

where U_t is the intrinsic velocity $\frac{d}{dt} Y_t$ at time t. This proves

Theorem 5.1. *The intrinsic velocity U_t of a homothermal acceleration wave satisfies*

$$U_t^2 = \frac{E_{Y_t}}{\varrho_R}, \tag{5.7}$$

where E_{Y_t} is the instantaneous stress-strain modulus at fixed present temperature corresponding to the histories $F^t(Y_t, \cdot)$ and $\vartheta^t(Y_t, \cdot)$ at the wave.

---

* Theorem 4.3 generalizes an observation long familiar in classical hydrodynamics. A proof for hyperelastic materials is given by Truesdell [8] with attributions to Duhem.

The jump in the acceleration is called the (kinematical) *amplitude* $a(t)$ of the acceleration wave:

$$a(t)=[\ddot{x}](t). \tag{5.8}$$

Let k_{Y_t} and φ_{Y_t} denote the values assumed at the wave by the moduli defined in (1.10) and (2.9),

$$k_{Y_t}=-\frac{\partial}{\partial g}\mathfrak{q}\left(F^t(Y_t,\cdot),\vartheta^t(Y_t,\cdot),g(Y_t,t)\right), \tag{5.9a}$$

$$\varphi_{Y_t}=D_F\mathfrak{q}\left(F^t(Y_t,\cdot),\vartheta^t(Y_t,\cdot),g(Y_t,t)\right), \tag{5.9b}$$

and let ϑ_{Y_t} be the temperature at the wave,

$$\vartheta_{Y_t}=\vartheta(Y_t,t). \tag{5.10}$$

We now wish to show that for a homothermal acceleration wave the numbers U_t, k_{Y_t}, α_{Y_t}, φ_{Y_t}, and ϑ_{Y_t} can be used to relate the kinematical amplitude $a(t)$ to the "thermal amplitude" $[\ddot{\vartheta}](t)$. We note first that since we here assume $[\dot{\vartheta}]=[\partial_X\vartheta]=0$, it follows from MAXWELL's theorem that

$$[\ddot{\vartheta}]=-U[\partial_X\dot{\vartheta}]=U^2[\partial_X^2\vartheta]. \tag{5.11}$$

The chain rule shown in (5.1) may be used to calculate $\partial_X q$ and $\partial_X\eta$ for any $X\neq Y_t$:

$$\begin{aligned}\partial_X q=&D_F\mathfrak{q}(F^t,\vartheta^t;g)\,\partial_X F+D_\vartheta\mathfrak{q}(F^t,\vartheta^t;g)\,\partial_X\vartheta+\delta_F\mathfrak{q}(F^t,\vartheta^t;g\,|\,\partial_X F_r^t)+\\&+\delta_\vartheta\mathfrak{q}(F^t,\vartheta^t;g\,|\,\partial_X\vartheta_r^t)+\partial_g\mathfrak{q}(F^t,\vartheta^t;g)\,\partial_X^2\vartheta,\end{aligned} \tag{5.12}$$

$$\begin{aligned}\partial_X\eta=&D_F\eta(F^t,\vartheta^t)\,\partial_X F+D_\vartheta\mathfrak{h}(F^t,\vartheta^t)\,\partial_X\vartheta+\\&+\delta_F\mathfrak{h}(F^t,\vartheta^t\,|\,\partial_X F_r^t)+\delta_\vartheta\mathfrak{h}(F^t,\vartheta^t\,|\,\partial_X\vartheta_r^t).\end{aligned} \tag{5.13}$$

When the argument starting at (5.1) and ending with (5.5) is applied to (5.12), one obtains

$$[\partial_X q]=\varphi_{Y_t}[\partial_X F]-k_{Y_t}[\partial_X^2\vartheta]. \tag{5.14}$$

The same argument applied to (5.13) yields

$$[\partial_X\eta]=D_F\mathfrak{h}\left(F^t(Y_t,\cdot),\vartheta^t(Y_t,\cdot)\right)[\partial_X F], \tag{5.15}$$

which, by virtue of Remark 3.4 and (5.3b), can be written in the form

$$[\partial_X\eta]=-\frac{1}{\varrho_R}\alpha_{Y_t}[\partial_X F]. \tag{5.16}$$

The equations (4.20) and (4.22) imply that the equation

$$[\partial_X q]=\varrho_R\vartheta_{Y_t}U_t[\partial_X\eta] \tag{5.17}$$

holds for all acceleration waves. Combining (5.17) with (5.16), we obtain the following interesting relation for homothermal waves:

$$[\partial_X q]=-U_t\alpha_{Y_t}\vartheta_{Y_t}[\partial_X F]. \tag{5.18}$$

Clearly, (5.14) and (5.18) yield the equation

$$k_{Y_t}[\partial_X^2\vartheta]=(\varphi_{Y_t}+U_t\alpha_{Y_t}\vartheta_{Y_t})[\partial_X F], \tag{5.19}$$

and this, when combined with (4.2) and (5.11), gives us

Theorem 5.2. *In a homothermal acceleration wave with intrinsic velocity* U_t

$$k_{Y_t}[\ddot{\vartheta}] = (\varphi_{Y_t} + U_t \alpha_{Y_t} \vartheta_{Y_t})\, a(t)\,. \tag{5.20}$$

Here ϑ_{Y_t} *is the temperature,* k_{Y_t} *the heat conduction modulus,* α_{Y_t} *the instantaneous stress-temperature modulus, and* φ_{Y_t} *the instantaneous heat-strain modulus, all evaluated at the wave.*

Starting with the observation that the familiar field equation governing the temperature field in a rigid linear heat conductor is parabolic, one might jump to the conclusion that the theory of homothermal waves is empty. Such a conclusion is, however, false. Indeed, RANKINE [11], in his analysis of shock structure in an inviscid ideal gas with Fourier heat conduction, obtained an exact solution which, in certain cases, exhibits a homothermal acceleration wave.*

6. Growth and Decay of Waves in Definite Conductors

We seek here a general formula for the rate of change of the amplitude of acceleration waves in materials for which the heat conduction modulus k_t is positive. By Theorem 4.1 such waves are homothermal.

We assume that the intrinsic velocity U_t is not zero. Remark 2.2 of II tells us that the amplitude a of an acceleration wave must obey the formula

$$2\frac{da}{dt} = \frac{a}{U_t}\frac{dU_t}{dt} + \frac{1}{\varrho_R}[\partial_t \partial_X T] - U_t^2[\partial_X \dot{F}]. \tag{6.1}$$

Since the derivation of this formula used only the hypothesis A-1, our smoothness assumption on b, and the law of balance of momentum, we can use (6.1) in our present discussion. Theorem 5.1 gives us an equation relating U_t to the motion. It remains for us to find a similar result for the quantity $[\partial_t \partial_X T]$.

Differentiating (5.1) with respect to t and putting $\Lambda^t = F^t, \vartheta^t$, we obtain the following formula which is valid whenever $X \neq Y_t$:

$$\begin{aligned}
\partial_t \partial_X T = {} & D_F \mathfrak{T}(\Lambda^t)\, \partial_X \dot{F} + D_F^2 \mathfrak{T}(\Lambda^t)\, \dot{F}\, \partial_X F + \delta_F D_F \mathfrak{T}(\Lambda^t | \dot{F}_r^t)\, \partial_X F + \\
& + D_\vartheta D_F \mathfrak{T}(\Lambda^t)\, \dot{\vartheta}\, \partial_X F + \delta_\vartheta D_F \mathfrak{T}(\Lambda^t | \dot{\vartheta}_r^t)\, \partial_X F + D_\vartheta \mathfrak{T}(\Lambda^t)\, \partial_X \dot{\vartheta} + \\
& + D_F D_\vartheta \mathfrak{T}(\Lambda^t)\, \dot{F}\, \partial_X \vartheta + \delta_F D_\vartheta \mathfrak{T}(\Lambda^t | \dot{F}_r^t)\, \partial_X \vartheta + D_\vartheta^2 \mathfrak{T}(\Lambda^t)\, \dot{\vartheta}\, \partial_X \vartheta + \\
& + \delta_\vartheta D_\vartheta \mathfrak{T}(\Lambda^t | \dot{\vartheta}_r^t)\, \partial_X \vartheta + D_F \delta_F \mathfrak{T}(\Lambda^t | \partial_X F_r^t)\, \dot{F} + \delta_F^2 \mathfrak{T}(\Lambda^t | \partial_X F_r^t, \dot{F}_r^t) + \\
& + D_\vartheta \delta_F \mathfrak{T}(\Lambda^t | \partial_X F_r^t)\, \dot{\vartheta} + \delta_\vartheta \delta_F \mathfrak{T}(\Lambda^t | \partial_X F_r^t, \dot{\vartheta}_r^t) + D_F \delta_\vartheta \mathfrak{T}(\Lambda^t | \partial_X \vartheta_r^t)\, \dot{F} + \\
& + \delta_F \delta_\vartheta \mathfrak{T}(\Lambda^t | \partial_X \vartheta_r^t, \dot{F}_r^t) + D_\vartheta \delta_\vartheta \mathfrak{T}(\Lambda^t | \partial_X \vartheta_r^t)\, \dot{\vartheta} + \delta_\vartheta^2 \mathfrak{T}(\Lambda^t | \partial_X \vartheta_r^t, \dot{\vartheta}_r^t) + \\
& + \frac{\partial}{\partial \tau} \delta_F \mathfrak{T}(\Lambda^t | \partial_X F_r^\tau)\Big|_{\tau=t} + \frac{\partial}{\partial \tau} \delta_\vartheta \mathfrak{T}(\Lambda^t | \partial_X \vartheta_r^\tau)\Big|_{\tau=t}.
\end{aligned} \tag{6.2}$$

It follows from A-1 and A-2 that Λ^t, as a point in $\mathscr{S}$, and $\partial_X F_r^t$, $\dot{F}_r^t$, $\partial_X \vartheta_r^t$, $\dot{\vartheta}_r^t$, as points in $\mathscr{H}_h$, are all continuous functions of X and t, even across Σ. Furthermore, for homothermal waves the present values $\dot{\vartheta}$ and $\partial_X \vartheta$ are also continuous

---

* RANKINE's solution is discussed by HAYES [12], p. 45. If in that discussion we take $u_p = 2u_2$, then the velocity, temperature, and temperature gradient are continuous everywhere, but the acceleration suffers a jump discontinuity at one point, Y_t. Throughout the region ahead of the wave, the velocity and temperature are constants independent of time and position.

across Σ. Therefore, by the continuity of the various "D" and "δ" derivatives of $\mathfrak{T}$ we have

$$[D_F \mathfrak{T}(\Lambda^t)\,\partial_X \dot{F}] = E_{Y_t}[\partial_X \dot{F}], \tag{6.3a}$$

$$[D_F^2 \mathfrak{T}(\Lambda^t)\,\dot{F}\,\partial_X F] = \widetilde{E}_{Y_t}[\dot{F}\,\partial_X F], \tag{6.3b}$$

$$[\delta_F D_F \mathfrak{T}(\Lambda^t|\dot{F}_r^t)\,\partial_X F + D_\vartheta D_F \mathfrak{T}(\Lambda^t)\,\dot{\vartheta}\,\partial_X F + \delta_\vartheta D_F \mathfrak{T}(\Lambda^t|\dot{\vartheta}_r^t)\,\partial_X F] = I_{Y_t}[\partial_X F], \tag{6.3c}$$

$$[D_\vartheta \mathfrak{T}(\Lambda^t)\,\partial_X \dot{\vartheta}] = \alpha_{Y_t}[\partial_X \dot{\vartheta}], \tag{6.3d}$$

$$[D_F D_\vartheta \mathfrak{T}(\Lambda^t)\,\dot{F}\,\partial_X \vartheta + D_F \delta_F \mathfrak{T}(\Lambda^t|\,\partial_X F_r^t)\,\dot{F} + D_F \delta_\vartheta \mathfrak{T}(\Lambda^t|\,\partial_X \vartheta_r^t)\,\dot{F}] = J_{Y_t}[\dot{F}], \tag{6.3e}$$

$$\begin{aligned}
&[\delta_F D_\vartheta \mathfrak{T}(\Lambda^t|\dot{F}_r^t)\,\partial_X \vartheta] = [D_\vartheta^2 \mathfrak{T}(\Lambda^t)\,\dot{\vartheta}\,\partial_X \vartheta] = [\delta_\vartheta D_\vartheta \mathfrak{T}(\Lambda^t|\dot{\vartheta}_r^t)\,\partial_X \vartheta] \\
&\quad = [D_\vartheta \delta_F \mathfrak{T}(\Lambda^t|\,\partial_X F_r^t)\,\dot{\vartheta}] = [\delta_\vartheta \delta_F \mathfrak{T}(\Lambda^t|\,\partial_X F_r^t, \dot{\vartheta}_r^t)] = [D_\vartheta \delta_\vartheta \mathfrak{T}(\Lambda^t|\,\partial_X \vartheta_r^t)\,\dot{\vartheta}] \\
&\quad = [\delta_\vartheta^2 \mathfrak{T}(\Lambda^t|\,\partial_X \vartheta_r^t, \dot{\vartheta}_r^t)] = [\delta_F^2 \mathfrak{T}(\Lambda^t|\,\partial_X F_r^t, \dot{F}_r^t)] = [\delta_F \delta_\vartheta \mathfrak{T}(\Lambda^t|\,\partial_X \vartheta_r^t, \dot{F}_r^t)] = 0.
\end{aligned} \tag{6.3f}$$

Here,

$$\widetilde{E}_{Y_t} = D_F^2 \mathfrak{T}(F^t(Y_t,\cdot), \vartheta^t(Y_t,\cdot)), \tag{6.4a}$$

$$\begin{aligned}
I_{Y_t} = {}& \delta_F D_F \mathfrak{T}(F^t(Y_t,\cdot), \vartheta^t(Y_t,\cdot)|\dot{F}_r^t(Y_t,\cdot)) + D_\vartheta D_F \mathfrak{T}(F^t(Y_t,\cdot), \vartheta^t(Y_t,\cdot))\,\dot{\vartheta}_{Y_t} + \\
& + \delta_\vartheta D_F \mathfrak{T}(F^t(Y_t,\cdot), \vartheta^t(Y_t,\cdot)|\dot{\vartheta}_r^t(Y_t,\cdot)),
\end{aligned} \tag{6.4b}$$

$$\begin{aligned}
J_{Y_t} = {}& D_F \delta_F \mathfrak{T}(F^t(Y_t,\cdot), \vartheta^t(Y_t,\cdot)|\,\partial_X F_r^t(Y_t,\cdot)) + \\
& + D_F D_\vartheta \mathfrak{T}(F^t(Y_t,\cdot), \vartheta^t(Y_t,\cdot))\,g_{Y_t} + D_F \delta_\vartheta \mathfrak{T}(F^t(Y_t,\cdot), \vartheta^t(Y_t,\cdot)|\,\partial_X \vartheta_r^t(Y_t,\cdot)),
\end{aligned} \tag{6.4c}$$

$$\dot{\vartheta}_{Y_t} = \partial_t \vartheta(X,t)|_{X=Y_t}, \qquad g_{Y_t} = \partial_X \vartheta(X,t)|_{X=Y_t}, \tag{6.4d}$$

and the numbers E_{Y_t} and α_{Y_t} are given by (5.3).

The equations (6.3) tell us the jump across Σ of all the terms in (6.2) except the last two. When we apply to the penultimate term the analysis used to derive equation (2.34) of II, we obtain

$$\left[\frac{\partial}{\partial \tau}\,\delta_F \mathfrak{T}(\Lambda^t|\,\partial_X F_r^\tau)\Big|_{\tau=t}\right] = G'_{Y_t}(0)\,[\partial_X F], \tag{6.5}$$

where

$$G'_{Y_t}(0) = \frac{\partial}{\partial s}\,\mathfrak{G}(F^t(Y_t,\cdot), \vartheta^t(Y_t,\cdot); s)\Big|_{s=0} \tag{6.6a}$$

is the initial slope of the stress-strain relaxation function (2.15) corresponding to the histories $F^t(Y_t,\cdot)$, $\vartheta^t(Y_t,\cdot)$. We may remark that by (2.14) the number E_{Y_t} of (5.3a) and (6.3a) equals the initial value of this same relaxation function:

$$E_{Y_t} = G_{Y_t}(0) = \mathfrak{G}(F^t(Y_t,\cdot), \vartheta^t(Y_t,\cdot); 0). \tag{6.6b}$$

To evaluate the last term in (6.2) we add to A-1 and A-2 one more hypothesis.

A-3*: $\partial_X \vartheta_r^t(X,\cdot)$ is a smooth function of X and t with respect to the norm $\|\cdot\|_h$ for all X and t.

It follows from A-3 and the smoothness of the functional $\delta_\vartheta \mathfrak{T}(\Lambda^t|\cdot)$ over $\mathscr{H}_h$ that

$$\left[\frac{\partial}{\partial \tau}\,\delta_\vartheta \mathfrak{T}(\Lambda^t|\,\partial_X \dot{\vartheta}_r^\tau)\Big|_{\tau=t}\right] = [\delta_\vartheta \mathfrak{T}(\Lambda^t|\,\partial_t \partial_X \vartheta_r^t)] = 0. \tag{6.7}$$

* In Remark 7.1 we observe that this assumption, which is natural for a homothermal wave, is implied by A-1 for a wave entering a homogeneous definite conductor at rest at constant temperature.

Combining (6.2) with (6.3), (6.5), and (6.7), we obtain

$$[\partial_t \partial_X T] = E_{Y_t}[\partial_X \dot{F}] + \tilde{E}_{Y_t}[\dot{F}\,\partial_X F] + \{I_{Y_t} + G'_{Y_t}(0)\}[\partial_X F] + \alpha_{Y_t}[\partial_X \dot{\vartheta}] + J_{Y_t}[\dot{F}]. \quad (6.8)$$

Using (5.11) and the assumption that $k_{Y_t} \neq 0$, we can write the conclusion of Theorem 5.2 in the form

$$[\partial_X \dot{\vartheta}] = -\frac{U_t\,\alpha_{Y_t}\,\vartheta_{Y_t} + \varphi_{Y_t}}{k_{Y_t}\,U_t}\,a(t). \quad (6.9)$$

This equation, (4.2), and Theorem 5.1 enable us to transform (6.8) into

$$\frac{1}{\varrho_R}[\partial_t \partial_X T] = U_t^2[\partial_X \dot{F}] + \frac{1}{\varrho_R}\tilde{E}_{Y_t}[\dot{F}\,\partial_X F] + \\ + \frac{1}{\varrho_R U_t}\left\{\frac{I_{Y_t} + G'_{Y_t}(0)}{U_t} - J_{Y_t} - \frac{U_t\,\alpha_{Y_t}^2\,\vartheta_{Y_t} + \varphi_{Y_t}\,\alpha_{Y_t}}{k_{Y_t}}\right\}a(t). \quad (6.10)$$

Substitution of this result into (6.1) yields

Theorem 6.1. *The amplitude $a = a(t)$ of an acceleration wave in a definite conductor obeys the relation*

$$\frac{da}{dt} = -\mu_t\,a + \beta_t[\dot{F}\,\partial_X F], \quad (6.11)$$

where

$$\mu_t = -\frac{1}{2U_t}\frac{dU_t}{dt} + \frac{1}{2\varrho_R U_t}\left\{J_{Y_t} - \frac{I_{Y_t} + G'_{Y_t}(0)}{U_t} + \frac{U_t\,\alpha_{Y_t}^2\,\vartheta_{Y_t} + \varphi_{Y_t}\,\alpha_{Y_t}}{k_{Y_t}}\right\}, \quad (6.12)$$

and

$$\beta_t = \frac{1}{2\varrho_R}\tilde{E}_{Y_t}; \quad (6.13)$$

U_t is the intrinsic velocity of the wave; ϑ_{Y_t} is the temperature at the wave; the numbers α_{Y_t}, k_{Y_t}, φ_{Y_t}, $\tilde{E}_{Y_t}$, I_{Y_t}, J_{Y_t}, and $G'_{Y_t}(0)$, are given by (5.3), (5.9), (6.4), *and* (6.6); *though these quantities depend on t, they are determined when $F^t(Y_t, \cdot)$, $\vartheta^t(Y_t, \cdot)$, $g(Y_t, t)$ and the functionals $\mathfrak{T}$ and $\mathfrak{q}$ are specified.*

The equation (6.11) has the form of equation (2.38) of II, which was derived under neglect of thermodynamic effects. Though the coefficient β_t is here no more complicated than the corresponding term in II, the coefficient μ_t involves several new features. In addition to the added quantity $\dfrac{U_t\,\alpha_{Y_t}^2\,\vartheta_{Y_t} - \varphi_{Y_t}\,\alpha_{Y_t}}{2\varrho_R\,k_{Y_t}\,U_t}$, which is smaller the greater the heat conduction coefficient k_{Y_t}, there are also new contributions to I_{Y_t} and J_{Y_t}. On comparing our present equations (6.4b) and (6.4c) with the equations (2.24) and (2.25) of II, we see that here we have $D_\vartheta D_F \mathfrak{T}(\Lambda^t)\,\dot{\vartheta} + \delta_\vartheta D_F \mathfrak{T}(\Lambda^t|\,\dot{\vartheta}_r^t)$ added to the old I_{Y_t} and $D_F D_\vartheta \mathfrak{T}(\Lambda^t)\,g + D_F \delta_\vartheta \mathfrak{T}(\Lambda^t|\,\partial_X \vartheta_r^t)$ added to the old J_{Y_t}. The new I_{Y_t} and J_{Y_t} can be expected to reduce to the old only if $\dot{\vartheta}$ and $\partial_X \vartheta$ are zero at the point Y_t for all times $\leqq t$.*

* If one considers a *perfect conductor of heat* or, more precisely, considers a sequence of materials with the same functional $\mathfrak{T}$, but with functionals $\mathfrak{q}_i$ with the property that $k_i \to \infty$ for each history, then one should find that in the limit of large k_i the only "reasonable" thermodynamic processes (*i.e.*, the only processes giving finite values for both $q(X_\alpha, t)$ and $q(X_\beta, t)$ in (1.21)) will be those for which ϑ is constant in space and time. *For such isothermal processes in perfect conductors our present eq.*(6.11) *reduces to eq.* (2.38) *of* II. Our present interest lies, however, in the more practical case of finite k.

When the sign of U_t is known, *i.e.*, when one can distinguish the region ahead of the wave from one behind the wave, (6.11) can be converted into a differential equation for $a(t)$ with coefficients determined by F^t, ϑ^t, g at the wave and $\partial_X F$, $\dot{F}$ ahead of the wave. For example, when U_t is positive, the numbers

$$(\partial_X F)_{Y_t^+} = \lim_{X\to Y_t^+} \partial_X F(X,t), \qquad \dot{F}_{Y_t^-} = \lim_{X\to Y_t^-} \partial_t F(X,t) \tag{6.14}$$

give the values of $\partial_X F$ and $\dot{F}$ just in front of the wave. Using the relation

$$[\dot{F}\,\partial_X F] = \dot{F}_{Y_t^+}[\partial_X F] + (\partial_X F)_{Y_t^+}[\dot{F}] + [\dot{F}][\partial_X F] \tag{6.15}$$

and the compatibility conditions (4.2), we can write

$$[\dot{F}\,\partial_X F] = \{U_t^{-2}\dot{F}_{Y_t^+} - U_t^{-1}(\partial_X F)_{Y_t^+}\}\,a(t) - U_t^{-3}\,a(t)^2, \tag{6.16}$$

and substituting this into (6.11) we obtain*

Remark 6.1. The amplitude of an acceleration wave with $U_t > 0$ propagating in a definite conductor obeys the differential equation

$$\frac{da}{dt} = \{\beta_t[U_t^{-2}\dot{F}_{Y_t^+} - U_t^{-1}(\partial_X F)_{Y_t^+}] - \mu_t\}\,a - U_t^{-3}\beta_t\,a^2 \tag{6.17}$$

with μ_t given by (6.12) and β_t by (6.13).

7. A Wave Entering a Definite Conductor at Rest in a Homogeneous State

We consider now an acceleration wave which since time $t=0$ has been propagating into a definite conductor which previously had been at rest in a fixed homogeneous configuration at a constant and uniform temperature. Thus, we suppose that $U_t > 0$ and that for $X > Y_t$ and $t \geqq 0$

$$x = x(X, t-s) = F_o X + X_o, \qquad 0 \leqq s < \infty, \tag{7.1a}$$

$$\vartheta = \vartheta(X, t-s) = \vartheta_o, \qquad 0 \leqq s < \infty. \tag{7.1b}$$

Before exhibiting the solution of (6.14) for such a wave, we make a remark whose proof is the same as the proof of Remark 4.1 of II.

Remark 7.1. In the present circumstances in which (7.1) holds with $U_t > 0$, the hypotheses A-2 and A-3, for $X > Y_0$ and $t > 0$, follow from A-1 and the assumption that the medium is a definite conductor.

Now, let $F_o^\dagger$ and $\vartheta_o^\dagger$ be the constant histories whose values are F_o and ϑ_o:

$$F_o^\dagger(s) = F_o, \qquad \vartheta_o^\dagger(s) = \vartheta_o, \qquad 0 \leqq s < \infty. \tag{7.2}$$

It follows from (7.1) that

$$\left.\begin{aligned} F^t(X,\cdot) &= F^t(Y_t,\cdot) = F_o^\dagger, \\ \vartheta^t(X,\cdot) &= \vartheta^t(Y_t,\cdot) = \vartheta_o^\dagger, \quad \text{for} \quad X \geqq Y_t \quad \text{and} \quad t \geqq 0, \end{aligned}\right\} \tag{7.3}$$

and hence that

$$\dot{F}^t(Y_t^+, s) = \partial_X F(Y_t^+, s) = \dot{\vartheta}^t(Y_t^+, s) = \partial_X \vartheta^t(Y_t^+, s) = 0, \qquad 0 \leqq s < \infty. \tag{7.4}$$

---

* The argument we use to pass from (6.11) to (6.14) is analogous to that used by VARLEY [9] in deriving his eq. (4.6) from his eq. (4.3).

The last two equalities of (7.4), when combined with the observation that a wave in a definite conductor is homothermal, yield zero for the numbers defined in (6.4d):

$$\dot{\vartheta}_{Y_t} = g_{Y_t} = 0. \tag{7.5}$$

Inserting (7.3) and (7.5) into (5.3), (5.9), (6.4a) and (6.6), we find that for $t \geqq 0$

$$\begin{aligned} &E_{Y_t} = G_{Y_t}(0) = G(0), \qquad G'_{Y_t}(0) = G'(0), \qquad \tilde{E}_{Y_t} = \tilde{E}, \\ &\alpha_{Y_t} = \alpha, \qquad k_{Y_t} = k, \qquad \varphi_{Y_t} = \varphi, \end{aligned} \tag{7.6}$$

where

$$\begin{aligned} G(0) &= \mathfrak{G}(F_o^\dagger, \vartheta_o^\dagger; 0) = D_F \mathfrak{T}(F_o^\dagger, \vartheta_o^\dagger), \\ G'(0) &= \frac{\partial}{\partial s} \mathfrak{G}(F_o^\dagger, \vartheta_o^\dagger; s)\Big|_{s=0}, \\ \tilde{E} &= D_F^2 \mathfrak{T}(F_o^\dagger, \vartheta_o^\dagger), \\ \alpha &= D_\vartheta \mathfrak{T}(F_o^\dagger, \vartheta_o^\dagger), \\ k &= -\frac{\partial}{\partial g} \mathfrak{q}(F_o^\dagger, \vartheta_o^\dagger; g)\Big|_{g=0}, \\ \varphi &= D_F \mathfrak{q}(F_o^\dagger, \vartheta_o^\dagger, 0), \end{aligned} \tag{7.7}$$

are all constants independent of t for $t \geqq 0$. Furthermore, (5.7) now reduces to

$$U_t = U = \sqrt{\frac{G(0)}{\varrho_R}} = \text{const.} \tag{7.8}$$

Since, by A-2, $\dot{F}_r^t$, $\partial_X F_r^t$, $\dot{\vartheta}_r^t$, and $\partial_X \vartheta_r^t$, as points in $\mathscr{H}_h$, are continuous functions of X at Y_t, it follows from (7.4) that, for $t \geqq 0$, $\dot{F}_r^t(Y_t, \cdot)$, $\partial_X F_r^t(Y_t, \cdot)$, $\dot{\vartheta}_r^t(Y_t, \cdot)$, and $\partial_X \vartheta_r^t(Y_t, \cdot)$ are all constant functions with value zero. On combining this observation with (7.5), we see that (6.4b) and (6.4c) here yield

$$I_{Y_t} = J_{Y_t} = 0, \qquad t \geqq 0. \tag{7.9}$$

It follows from the equations (6.12), (6.13), (7.4), and (7.6)—(7.10) that whenever (7.1) holds, (6.17) takes the form

$$\frac{da}{dt} = -\mu a + \frac{\mu}{\lambda} a^2, \tag{7.10}$$

where

$$\begin{aligned} \mu &= -\frac{G'(0)}{2G(0)} + \frac{\alpha}{2k\,\varrho_R}\left\{\alpha\vartheta_o + \frac{\varphi}{U}\right\} = \text{const.} \\ \lambda &= -\frac{2U\,G(0)\,\mu}{\tilde{E}} = \frac{U\,G'(0)}{\tilde{E}} - \frac{U^3\alpha}{k\tilde{E}}\left\{\alpha\vartheta_o + \frac{\varphi}{U}\right\} = \text{const.} \end{aligned} \tag{7.11}$$

Integration of (7.11) yields

Theorem 7.1. *Consider an acceleration wave propagating into a definite conductor of infinite extent, which previous to the arrival of the wave had always been at rest with its deformation gradient F_o and temperature ϑ_o constant in space and time;* i.e., *suppose that $U_t > 0$ and that (7.1) holds for $X \geqq Y_t$, $t \geqq 0$. Then, for $t \geqq 0$ the intrinsic velocity U_t is a constant given by (7.8), and the amplitude a is the following explicit function of t:*

$$a(t) = \frac{\lambda}{\left(\frac{\lambda}{a(0)} - 1\right)e^{\mu t} + 1}. \tag{7.12}$$

For a given material, μ and λ are constants depending on only F_o and ϑ_o as shown in (7.11), (7.7) and (7.8).

For most materials we expect to have $\varphi=0$, and, hence,

$$\mu=-\frac{G'(0)}{2G(0)}+\frac{\alpha^2\vartheta_o}{2k\,\varrho_R},\qquad \lambda=\frac{U\,G'(0)}{\tilde{E}}-\frac{U^3\alpha^2\vartheta_o}{k\tilde{E}}. \tag{7.13}$$

Indeed, if equation (3.10) holds for arbitrary F^t, ϑ^t, we must have

$$D_F\mathfrak{q}(F^t,\vartheta^t,0)=0, \tag{7.14}$$

and, by $(7.7)_6$, $\varphi=0$. In Remark 38 of [3]* it is pointed out that (3.10) holds for all F^t, ϑ^t whenever the symmetry group of the material contains the inversion transformation (*i.e.*, a reflection through a center of symmetry).** We summarize in

Remark 7.2. Whenever the material is such that $g=0$ implies $q=0$, and we expect this to be the case in applications, the coefficients λ and μ in (7.12) must take the simple forms (7.13).

In § 8 of [4] it is shown that our thermodynamic postulates require $G(0)\geqq 0$. Since we here assume $U\neq 0$, $G(0)=0$ is impossible; hence we have $G(0)>0$. We take as physically plausible the assumption that $G'(0)$ be not positive. The following is a list of the assumptions made so far about the material parameters occurring in (7.13):

$$\varrho_R>0,\qquad \vartheta_o>0,\qquad k>0,\qquad G(0)>0,\qquad G'(0)\leqq 0. \tag{7.15}$$

We know of no general physical principle restricting the sign of $\tilde{E}$ and α.

When $\tilde{E}\neq 0$, the following remark is implied by Theorem 7.1 and Remark 7.2 and generalizes Theorem 4.2 of II.

Remark 7.3. Let the hypotheses of Theorem 7.1 and Remark 7.2 hold, and suppose that in addition to (7.15)

$$\tilde{E}\neq 0 \quad\text{and either}\quad G'(0)\neq 0 \quad\text{or}\quad \frac{\alpha}{k}\neq 0. \tag{7.16}$$

If *either* $|a(0)|<|\lambda|$ *or* $\operatorname{sgn} a(0)=\operatorname{sgn}\tilde{E}$, then $a(t)\to 0$ monotonically as $t\to\infty$. If $a(0)=\lambda$, then $a(t)\equiv U(0)$. On the other hand, if *both* $|a(0)|>|\lambda|$ *and* $\operatorname{sgn} a(0)=-\operatorname{sgn}\tilde{E}$, then $|a(t)|\to\infty$ monotonically in a finite time t_∞ given by

$$t_\infty=-\frac{1}{\mu}\ln\left(1-\frac{\lambda}{a(0)}\right). \tag{7.17}$$

Our present Theorem 7.1 should be set alongside of Theorem 4.1 of II. We have deliberately used a notation which makes it apparent that the functions $a(t)$ in (7.12)

* The argument used there is that given by COLEMAN & NOLL in § 3 of [10].

** On p. 41 of [3] it is observed that when $F^t(s)=$const. and $\vartheta^t(s)=$const. it follows from (3.7) that $\sigma=0$ in (3.8) and hence that $q\cdot g\geqq 0$. Using an argument given in § 3 of [10], it can then be shown that (3.10) holds when F^t and ϑ^t are constant histories:

$$\mathfrak{q}(F_o^\dagger,\vartheta_o^\dagger,0)=0.$$

Although this equation, derived from general thermodynamic principles, must hold for all materials, it is not quite strong enough to give us $\varphi=0$.

above and in (4.12) of II have the same form. The theory of II, which neglects all thermal influences, gives, however, the following values for μ and λ:*

$$\mu = -\frac{G'(0)}{2G(0)}, \qquad \lambda = \frac{U\,G'(0)}{\tilde{E}}. \tag{7.18}$$

These values differ from those given here in (7.13) by terms which, for fixed values of $G(0)$, ϑ_o, $\tilde{E}$ and ϱ_R, are proportional to α^2/k. We note that since ϑ_o, k, and ϱ_R are positive in a definite conductor, the contribution $\frac{\alpha^2\vartheta_o}{2k\,\varrho_R}$ of "thermal dissipation" to the "damping coefficient" μ is never negative. We now call the number $|\lambda|$ defined by

$$|\lambda| = \left|\frac{G'(0)\,U}{\tilde{E}} - \frac{U^3\alpha^2\vartheta_o}{k\tilde{E}}\right| \tag{7.19}$$

the *critical amplitude*. This positive number plays here precisely the role that the number $\left|\frac{G'(0)\,U}{\tilde{E}}\right|$ played in § 4 of II (compare Remark 7.2 above with Theorem 4.2 of II). It follows from (7.15) that the critical amplitude of our present theory cannot be less than $\left|\frac{G'(0)\,U}{\tilde{E}}\right|$.

The forms assumed by (7.12) in the special cases $\frac{\alpha}{k} = G'(0) = 0$ and $\frac{\alpha}{k} = \tilde{E} = 0$ are discussed in Remarks 4.2 and 4.3 of II. The following remark covers a new and particularly interesting special case.

Remark 7.4. If the hypotheses of Theorem 7.1 and Remark 7.2 hold and, in addition, the material is such that

$$G'(0) = 0, \qquad \tilde{E} = 0, \tag{7.20}$$

then the amplitude has the following simple time-dependence:

$$a(t) = a(0)\exp\left\{-\frac{\alpha^2\vartheta_o\,t}{2k\,\varrho_R}\right\}. \tag{7.21}$$

This remark is, of course, applicable when the material is a perfectly elastic heat conductor with a linear stress-strain law.

For $\tilde{E}$ to be zero it is not necessary that the stress-strain law be linear. For, as we saw in Remark 6.1 of II, $\tilde{E}$ is zero at a shear acceleration wave propagating along a principal axis of strain into an isotropic material at equilibrium. Hence we have

Remark 7.5. Equation (7.21) is obeyed by the amplitude of a shear acceleration wave propagating along a principal axis of strain into an isotropic, perfectly elastic, definite conductor at rest in a homogeneous configuration with uniform temperature. This is true even if the stress-strain law for shear is nonlinear.

When the present theory is applied to longitudinal waves** in isotropic or transversely isotropic materials, the density ϱ is determined by F through the relation $\varrho(X,t) = F(X,t)^{-1}\varrho_R$, from which, by (4.2) and (5.8), follows the equation $a = \varrho_R\varrho^{-2}U[\dot{\varrho}]$. A wave is called *compressive* if $U(t)[\dot{\varrho}](t) > 0$ and *expansive*

* II, eq. (4.11).

** See § 6 of I and § 5 of II; for a longitudinal wave the motion is assumed to be always *plane longitudinal.*

if $U(t)[\dot{\varrho}](t)<0$. Let us assume that $\widetilde{E}\neq 0$ and put

$$\lambda_* = \frac{G'(0)\,\varrho_R}{\widetilde{E}\,F_0^2} - \frac{\alpha^2 G(0)\,\vartheta_o}{k\,\widetilde{E}\,F_0^2}. \tag{7.22}$$

The next remark, following from our Remark 7.3 and generalizing to definite conductors Remark 5.2 of II, tells us that $|\lambda_*|$ may be called the *critical jump in the density rate.*

Remark 7.6. Consider a longitudinal acceleration wave obeying the hypotheses of Theorem 7.1 and Remark 7.2 and propagating into a material obeying (7.15). If the wave is either initially compressive with $\widetilde{E}>0$ *or* initially expansive with $\widetilde{E}<0$, *or* if $|[\dot{\varrho}](0)|<|\lambda_*|$, then $[\dot{\varrho}](t)\to 0$ monotonically as $t\to\infty$. If $[\dot{\varrho}](0)=\lambda_*$, then $[\dot{\varrho}](t)\equiv[\dot{\varrho}](0)$. But, if $|[\dot{\varrho}](0)|>|\lambda_*|$, *and* the wave is either initially compressive with $\widetilde{E}<0$ or initially expansive with $\widetilde{E}>0$, then $|[\dot{\varrho}]|\to\infty$ monotonically in the finite time

$$t_\infty = \left\{\frac{G'(0)}{2G(0)} - \frac{\alpha^2\,\vartheta}{2k\,\varrho_R}\right\}^{-1} \ln\left(1-\left|\frac{\lambda_*}{[\dot{\varrho}](0)}\right|\right). \tag{7.23}$$

According to (7.12) and (7.13), if one lets $k\to 0$ with $G(0)$, $G'(0)$, ϑ_0, $\widetilde{E}$, ϱ_R held fixed at any values compatible with (7.15) and α fixed at any nonzero value, then one finds that $a(t)\to 0$ for all $t>0$. We interpret this as meaning that when $k=0$ a homothermal wave is impossible in a homogeneous (non-conducting) medium at rest with $\alpha\neq 0$, a conclusion which, in view of Theorems 4.2 and 4.3, is not surprising. The special case of $k=0$ is treated in Sections 10 and 11, assuming, in accordance with Theorem 4.3, that the wave is homentropic.

One might suspect that the presence of heat conduction causes disturbances to be propagated ahead of the wave and by such reasoning conclude that the region in front of a wave in a definite conductor cannot be at rest. However, for an inviscid ideal gas with Fourier heat conduction, we do have an example of an acceleration wave steadily propagating into a region at rest. The example is furnished by RANKINE's solution mentioned at the end of Section 5.

8. Entropy as an Independent Variable

The remainder of this article is concerned with homentropic waves; since the theory of such waves is simpler if η^t is taken as an independent variable rather than ϑ^t, we now replace the equations (1.6a)—(1.6c) by the following constitutive equations:

$$\varepsilon(t)=\hat{\mathfrak{e}}(F^t,\eta^t), \tag{8.1a}$$

$$T(t)=\hat{\mathfrak{T}}(F^t,\eta^t), \tag{8.1b}$$

$$\vartheta(t)=\hat{\mathfrak{t}}(F^t,\eta^t). \tag{8.1c}$$

We assume that the functionals $\hat{\mathfrak{e}}$, $\hat{\mathfrak{T}}$, $\hat{\mathfrak{t}}$ enjoy, throughout their common domain $\hat{\mathscr{D}}$, all the smoothness properties which the principle of fading memory gives to $\mathfrak{p}$, $\mathfrak{T}$, $\mathfrak{h}$. In particular, we assume that $\hat{\mathscr{D}}$ is an open subset of $\mathscr{S}$ and that if we put $\Lambda^t=F^t,\eta^t$ then (1.9) holds with $N=3$ when $\mathfrak{f}=\hat{\mathfrak{e}}$ and with $N=2$ when $\mathfrak{f}=\hat{\mathfrak{T}}$ or $\hat{\mathfrak{t}}$.

Clearly, by replacing ϑ by η and ϑ_r^t by η_r^t in (2.3) and (2.4), we can define operators D_η, δ_η, D_F, δ_F which operate on $\mathfrak{f}(F^t,\eta^t)$ to give $D_\eta\mathfrak{f}(F^t,\eta^t)$, *etc.* Each of these operators commutes with the others.

In view of (1.3), the existence of the equations (8.1a)—(8.1c) is implied by the equations (1.6a)—(1.6c) if, and only if, the functional transformation

$$(F^t, \vartheta^t) \to (F^t, \eta^t), \tag{8.2a}$$

defined by the functional $\mathfrak{h}$ with

$$\eta^t(s) = \mathfrak{h}(F^{t-s}, \vartheta^{t-s}), \tag{8.2b}$$

is a *one-to-one* mapping of $\mathscr{D}$ onto $\hat{\mathscr{D}}$. When this is the case (8.1c) is the "inverse" of (1.6c),* and the following relation holds:

$$D_\eta \hat{\mathfrak{t}}(F^t, \eta^t) = [D_\vartheta \mathfrak{h}(F^t, \vartheta^t)]^{-1}. \tag{8.3}$$

Assuming the invertibility of the functional transformation (8.2), COLEMAN** has shown that the stress relation (3.2) and the entropy relation (3.3) imply that $\hat{\mathfrak{e}}$ determines $\hat{\mathfrak{T}}$ and $\hat{\mathfrak{t}}$ through the relations

$$\hat{\mathfrak{T}} = \varrho_R D_F \hat{\mathfrak{e}}, \tag{8.4}$$

$$\hat{\mathfrak{t}} = D_\eta \hat{\mathfrak{e}}. \tag{8.5}$$

It follows from (8.4) and (8.5) that

$$D_\eta \hat{\mathfrak{T}} = \varrho_R D_F \hat{\mathfrak{t}}. \tag{8.6}$$

In analogy to equations (2.6)—(2.8), for each pair of histories F^t, η^t, we define the corresponding *instantaneous stress-strain modulus* $\hat{E}_t$ *at fixed present entropy instantaneous second-order stress-strain modulus* $\hat{\hat{E}}$ *at fixed present entropy*, and *instantaneous stress-entropy modulus* $\hat{\alpha}_t$ *at fixed present strain*, through the relations

$$\hat{E}_t = D_F \hat{\mathfrak{T}}(F^t, \eta^t), \tag{8.7}$$

$$\hat{\hat{E}}_t = D_F^2 \hat{\mathfrak{T}}(F^t, \eta^t), \tag{8.8}$$

$$\hat{\alpha}_t = D_\eta \hat{\mathfrak{T}}(F^t, \eta^t). \tag{8.9}$$

According to (8.6) an alternative formula for $\hat{\alpha}_t$ is

$$\hat{\alpha}_t = \varrho_R D_F \hat{\mathfrak{t}}(F^t, \eta^t). \tag{8.10}$$

It follows from (1.6) and (1.3) that ε is given by a functional $\mathfrak{e}$ of F^t, ϑ^t:

$$\varepsilon = \mathfrak{e}(F^t, \vartheta^t) = \mathfrak{p}(F^t, \vartheta^t) + \vartheta(t)\, \mathfrak{h}(F^t, \vartheta^t). \tag{8.11}$$

Let us assume that $D_\vartheta \mathfrak{e}$ is not zero. The number $\varkappa_t$ defined by

$$\varkappa_t = D_\vartheta \mathfrak{e}(F^t, \vartheta^t) \tag{8.12}$$

may be called the *instantaneous heat capacity at fixed present strain* corresponding to F^t, ϑ^t; it measures the rate of change of internal energy with temperature at fixed strain in thermal impulses superimposed on F^t, ϑ^t at time t. From (8.12), (8.11), (3.3), and (8.3) we get

$$\varkappa_t = \vartheta(t)\, D_\vartheta \mathfrak{h}(F^t, \vartheta^t) = \vartheta(t)\, [D_\eta \hat{\mathfrak{t}}(F^t, \eta^t)]^{-1}. \tag{8.13}$$

This result may be used to relate the moduli $\hat{\alpha}_t$ and $\hat{E}_t$ to α_t and E_t, respectively.

* [3], § 11.
** [3], Theorem 6.

Using a self-evident extension of a familiar chain rule for composite functions, we have*

$$D_\eta \hat{\mathfrak{T}}(F^t, \eta^t) = D_\vartheta \mathfrak{T}(F^t, \vartheta^t)\, D_\eta \hat{\mathfrak{t}}(F^t, \eta^t), \tag{8.14}$$

and on combining this with (8.9), (2.8), and (8.13) we obtain

Remark 8.1. Let the pair F^t, ϑ^t be in $\mathscr{D}$, and let η^t be given by (8.2). Then the corresponding stress-entropy and stress-temperature moduli, (2.8) and (8.9), are related by the formula

$$\hat{\alpha}_t = \frac{\vartheta(t)}{\varkappa_t}\, \alpha_t \tag{8.15}$$

in which $\varkappa_t$ is the instantaneous heat capacity (8.12) and $\vartheta(t)$ the present temperature $\vartheta^t(0)$.

Another chain rule similar to (8.14) is**

$$D_F \hat{\mathfrak{T}}(F^t, \eta^t) = D_F \mathfrak{T}(F^t, \vartheta^t) + D_\vartheta \mathfrak{T}(F^t, \vartheta^t)\, D_F \hat{\mathfrak{t}}(F^t, \eta^t); \tag{8.16}$$

on substituting into this the equations (8.7), (2.6), (2.8), and (8.10) we find that

$$\hat{E}_t = E_t + \frac{\alpha_t \hat{\alpha}_t}{\varrho_R}. \tag{8.17}$$

Equations (8.15) and (8.17) yield

Remark 8.2. Whenever the pair F^t, ϑ^t is in $\mathscr{D}$ and η^t is given by (8.2), the instantaneous stress-strain moduli defined in (2.6) and (8.7) are related by the formula

$$\hat{E}_t = E_t + \frac{\alpha_t^2\, \vartheta(t)}{\varrho_R\, \varkappa_t} \tag{8.18}$$

with α_t and $\varkappa_t$ given by (2.8) and (8.12) and $\vartheta(t) = \vartheta^t(0)$.

In practical applications we expect to have $\varkappa_t > 0$, which, by (8.18), implies $\hat{E}_t \geqq E_t$.

When the argument giving us equations (2.10), (2.11), and (2.12) is applied to the functional $\delta_F \hat{\mathfrak{T}}$, we see that for each pair F^t, η^t there exists a function $\hat{G}'_t$, whose dependence on F^t, η^t may be expressed by writing

$$\hat{G}'_t(s) = \hat{\mathfrak{G}}'(F^t, \eta^t; s), \tag{8.19}$$

and which has the property that for each f in $\mathscr{H}_h$

$$\delta_F \hat{\mathfrak{T}}(F^t, \eta^t | f) = \int_0^\infty \hat{G}'_t(s)\, f(s)\, ds; \tag{8.20}$$

furthermore, G'_t is such that the function $G'_t h^{-2}$ is in $\mathscr{H}_h$. We assume that $\hat{\mathfrak{G}}'$ enjoys all the regularity properties postulated for $\mathfrak{G}'$ after equation (2.12). The solution $\hat{G}_t$ of

$$\frac{d}{ds}\hat{G}_t(s) = \hat{G}'_t(s), \qquad \hat{G}_t(0) = \hat{E}_t = D_F \hat{\mathfrak{T}}(F^t, \eta^t), \tag{8.21}$$

---

* *Cf.* [3], eq. (11.5a).
** *Cf.* [1], eq. (11.6).

may be called the *stress-strain relaxation function* corresponding to F^t, η^t;* to express the dependence of this function on F^t, η^t we write

$$\hat{G}_t(s) = \hat{\mathfrak{G}}(F^t, \eta^t; s). \tag{8.22}$$

In analogy to Remark 2.1 we have

Remark 8.3. If $F(\tau)$, $F^*(\tau)$, and $\eta(\tau)$, with $-\infty < \tau < \infty$, are such that for each t the pairs F^t, η^t and F^{*t}, η^t are in $\hat{\mathscr{D}}$ and the number ϵ defined by

$$\epsilon = \sup_{-\infty<\tau<\infty} |\gamma(\tau)|, \qquad \gamma(\tau) = F^*(\tau) - F(\tau) \tag{8.23}$$

is small, then we have, for all t,

$$T^*(t) = T(t) + \hat{G}_t(0)\,\gamma(t) + \int_0^\infty \hat{G}_t'(s)\,\gamma(t-s)\,ds + o(\epsilon), \tag{8.24}$$

where

$$T(t) = \hat{\mathfrak{T}}(F^t, \eta^t), \qquad T^*(t) = \hat{\mathfrak{T}}(F^{*t}, \eta^t), \tag{8.25}$$

and $\hat{G}_t$ is the stress-strain relaxation corresponding to F^t, η^t.

Considering now the functional $\hat{\mathfrak{e}}$, we note that for each pair F^t, η^t there exists a function $\hat{H}_t'$ such that $H_t' h^{-2}$ is in $\mathscr{H}_h$ and

$$\delta_F \hat{\mathfrak{e}}(F^t, \eta^t | f) = \int_0^\infty \hat{H}_t'(s)\, f(s)\, ds \tag{8.26}$$

for each f in $\mathscr{H}_h$. We call the solution $\hat{H}_t$ of

$$\frac{d}{ds}\hat{H}_t(s) = \hat{H}_t'(s), \tag{8.27}$$

$$\hat{H}_t(0) = D_F \hat{\mathfrak{e}}(F^t, \eta^t), \tag{8.28}$$

the *energy-strain relaxation function* corresponding to F^t, η^t.** To indicate the dependence of this function on F^t, η^t, we write

$$\hat{H}_t(s) = \hat{\mathfrak{H}}(F^t, \eta^t; s). \tag{8.29}$$

Since it follows from (8.28) and the stress relation (8.4) that

$$\hat{H}_t(0) = \frac{1}{\varrho_R}\mathfrak{T}(F^t, \eta^t), \tag{8.30}$$

the present counterpart of Remarks 2.1 and 8.3 is

Remark 8.4. If $F(\tau)$, $F^*(\tau)$ and $\eta(\tau)$ obey the hypothesis of Remark 8.3, then for each t we have

$$\varepsilon^*(t) = \varepsilon(t) + \frac{1}{\varrho_R} T(t)\,\gamma(t) + \int_0^\infty \hat{H}_t'(s)\,\gamma(t-s)\,ds + o(\epsilon), \tag{8.31}$$

---

* We should perhaps call $\hat{G}_t$ the *stress-strain relaxation function at fixed entropy history* corresponding to F^t and η^t. [See the footnote to the sentence following eq. (2.15).]

** *I.e.*, the *energy-strain relaxation function at fixed entropy history* corresponding to F^t, η^t.

where $\in$ and γ are given by (8.23), $T(t)$ by $(8.25)_1$,

$$\varepsilon(t) = \hat{\mathfrak{e}}(F^t, \eta^t), \qquad \varepsilon^*(t) = \hat{\mathfrak{e}}(F^{*t}, \eta^t), \tag{8.32}$$

and $\hat{H}_t$ is the energy-strain relaxation function corresponding to F^t, η^t.

Using (1.3), one can easily show that the *internal dissipation* σ of (3.5) is also given by*

$$\sigma = \frac{1}{\vartheta}\left[\frac{1}{\varrho_R} T\dot{F} + \vartheta\dot{\eta} - \dot{\varepsilon}\right], \tag{8.33}$$

which, upon use of (8.4), can be written

$$\sigma = \frac{1}{\vartheta(t)}\left[D_F \hat{\mathfrak{e}}(F^t, \eta^t)\dot{F} + D_\eta \hat{\mathfrak{e}}(F^t, \eta^t)\dot{\eta} - \frac{d}{dt}\hat{\mathfrak{e}}(F^t, \eta^t)\right]. \tag{8.34}$$

Hence, the internal dissipation inequality (3.6) may be regarded equally well as a restriction on $\mathfrak{p}$ or $\hat{\mathfrak{e}}$. It follows from the internal dissipation inequality that $\delta_F \mathfrak{p}$, $\delta_\vartheta \mathfrak{p}$, $\delta_F \hat{\mathfrak{e}}$, and $\delta_\eta \hat{\mathfrak{e}}$ must vanish at "equilibrium"; *i.e.*, whenever F^t, ϑ^t, η^t are the constant functions $F^\dagger, \vartheta^\dagger, \eta^\dagger$ defined by

$$F^\dagger(s) \equiv F(t), \qquad \vartheta^\dagger(s) \equiv \vartheta(t), \qquad \eta^\dagger(s) \equiv \eta(t), \qquad 0 \leqq s < \infty, \tag{8.35}$$

we have

$$\delta_F \mathfrak{p}(F^\dagger, \vartheta^\dagger | f) = \delta_\vartheta \mathfrak{p}(F^\dagger, \vartheta^\dagger | f) = \delta_F \hat{\mathfrak{e}}(F^\dagger, \eta^\dagger | f) = \delta_\eta \hat{\mathfrak{e}}(F^\dagger, \eta^\dagger | f) = 0, \tag{8.36}$$

for every function f in $\mathscr{H}_h$.**

It follows from (8.26) that the equation $(8.36)_3$ is equivalent to the following nontrivial remark.

Remark 8.5. If $\hat{H}$ is the energy-strain relaxation function corresponding to the constant histories $F^\dagger$ and $\eta^\dagger$, then $\hat{H}$ is itself a constant function, *i.e.*,

$$\hat{H}'(s) = \frac{d}{ds}\hat{H}(s) = 0 \quad \text{for} \quad 0 \leqq s < \infty.\text{***}$$

9. The Velocity of Homentropic Waves

When discussing homentropic waves, it is awkward to define an acceleration wave by the properties A-1 and A-2 of § 4. We therefore now replace ϑ by η in our previous definition and say that Y_t, with $-\infty < t < \infty$, is an *acceleration wave* if the fields $x(X, t)$ and $\eta(X, t)$ have the following properties:

---

* [3], eq. (11.14).

** This corollary to Theorem 3 and Remark 22 of [3] is expressed in eqs. (8.22) and (11.24) of [3]. Eq. (9.3) of [4] tells us that the functionals $\delta_{F_d^t}\mathfrak{p}$, $\delta_{\vartheta_d^t}\mathfrak{p}$, $\delta_{F_d^t}\hat{\mathfrak{e}}$, $\delta_{\eta_d^t}\hat{\mathfrak{e}}$ of [3] correspond to our present $\delta_F \mathfrak{p}$, $\delta_\vartheta \mathfrak{p}$, $\delta_F \mathfrak{e}$, $\delta_\eta \mathfrak{e}$.

*** Our present Remarks 8.4 and 8.5 imply that when $F^t = F^\dagger$, $\eta^t = \eta^\dagger$, we have

$$\varepsilon^*(t) = \varepsilon(t) + \frac{1}{\varrho_R} T(t)\,\gamma(t) + o(\in), \tag{8.37}$$

which equation supplies immediately a proof of the assertion that "at equilibrium", (8.4) reduces the familiar classical relation $T = \varrho_R \frac{\partial}{\partial F}\varepsilon(F, \eta)$. For the original proof see Remarks 11 and 23 of [3].

A-1′: $x, \dot{x}, F$, and η are continuous functions of X and t jointly for all X and t, while $\ddot{x}, \dot{F}, \partial_X F, \dot{\eta}, \partial_X \eta$ and also $\dddot{x}, \ddot{F}, \partial_X \dot{F}, \partial_X^2 F, \ddot{\eta}, \partial_X \dot{\eta}$, and $\partial_X^2 \eta$ have (at most) jump discontinuities across Σ but are continuous in X and t everywhere else.

A-2′*: $F_r^t(X, \cdot)$ and $\eta_r^t(X, \cdot)$ are smooth functions of X and t with respect to the norm $\|\cdot\|_h$ for all X and t.

We also assume that for each X and t, the pair $F^t(X, \cdot)$, $\eta^t(X, \cdot)$ is interior to $\hat{\mathscr{D}}$ and that b and r are assigned in such a way that $b, r, \dot{b}$ and $\partial_X r$ are continuous in X for all t.

Using the constitutive equations (8.1a)—(8.1b) and (1.6d), it is easy to verify that all the conclusions reached in Section 4 using A-1 and A-2 remain valid here.** Hence, the present definition of an acceleration wave is consistent with that of Section 4.

For example, A-1′, A-2′, and the smoothness of $\hat{\mathrm{e}}$ imply that for $X \neq Y_t$ we can write

$$\dot{\varepsilon} = D_F \hat{\mathrm{e}}(F^t, \eta^t) \dot{F} + D_\eta \hat{\mathrm{e}}(F^t, \eta^t) \dot{\eta} + \delta_F \hat{\mathrm{e}}(F^t, \eta^t | \dot{F}_r^t) + \delta_\eta \hat{\mathrm{e}}(F^t, \eta^t | \dot{\eta}_r^t). \tag{9.1}$$

By the continuity of $D_F \hat{\mathrm{e}}$, $D_\vartheta \hat{\mathrm{e}}$, $\delta_\eta \hat{\mathrm{e}}$, $\delta_F \hat{\mathrm{e}}$, and the fact that A-1′ and A-2′ imply that the function pair F^t, η^t, as a point in $\mathscr{S}$, and the functions $\dot{F}_r^t$ and $\dot{\eta}_r^t$, as points in $\mathscr{H}_h$ all depend continuously on X and t across Σ, we have

$$[\dot{\varepsilon}] = D_F \hat{\mathrm{e}}(F^t, \eta^t) [\dot{F}] + D_\eta \hat{\mathrm{e}}(F^t, \eta^t) [\dot{\eta}]. \tag{9.2}$$

Inserting (8.4) and (8.5) into (9.2), we obtain (4.14)***, from which we may read off (4.18) for homentropic waves, *i.e.*, waves obeying (4.16). The balance laws (1.1) and (1.2) give us again Remark 4.1. Insertion of (4.8) into (4.14) yields (4.22), the crucial step in proving Theorem 4.3. Thus, *under our present definition every acceleration wave with $U \neq 0$ in a non-conductor is again homentropic and, therefore, obeys*

$$[\partial_X \eta] = [\dot{\eta}] = 0, \tag{9.3}$$

$$T[\dot{F}] = \varrho_R [\dot{\varepsilon}]. \tag{9.4}$$

We now use A-1′, A-2′, and the smoothness of $\hat{\mathfrak{T}}$ to calculate the velocity of homentropic waves. Clearly, for $X \neq Y_t$ we have

$$\begin{aligned} \partial_X T = D_F \hat{\mathfrak{T}}(F^t, \eta^t)\, \partial_X F + D_\eta \hat{\mathfrak{T}}(F^t, \eta^t)\, \partial_X \eta + \\ + \delta_F \hat{\mathfrak{T}}(F^t, \eta^t | \partial_X F_r^t) + \delta_\eta \hat{\mathfrak{T}}(F^t, \eta^t | \partial_X \eta_r^t). \end{aligned} \tag{9.5}$$

Applying to (9.5) the argument used to derive (5.2) from (5.1), we here find that

$$[\partial_X T] = D_F \hat{\mathfrak{T}}(F^t, \eta^t) [\partial_X F] + D_\eta \hat{\mathfrak{T}}(F^t, \eta^t) [\partial_X \eta]. \tag{9.6}$$

---

* In Remark 11.1 we observe that for a wave entering a homogeneous non-conductor at rest at constant entropy the assumption of A-1′ implies A-2′.

** This includes Remark 4.1 and Theorems 4.1—4.3.

*** We note that here the smoothness of $\hat{\mathfrak{t}}$ gives us $[\vartheta] = 0$, and hence, by (1.3), we again have (4.13). Combination of (4.13) with (4.14) yields (4.12). These are the main steps in rederiving (4.17) for homothermal waves. The other part of Theorem 4.1 and Theorem 4.3 are reproven above. Once we have $[\vartheta] = 0$, we can reuse the old proof of Theorem 4.2.

Therefore, if we put

$$\widehat{E}_{Y_t} = D_F \widehat{\mathfrak{T}}(F^t(Y_t,\cdot), \eta^t(Y_t,\cdot)) \tag{9.7}$$

and note that $[\partial_X \eta] = 0$ for homentropic waves, we have

$$[\partial_X T] = \widehat{E}_{Y_t} [\partial_X F], \tag{9.8}$$

which equation, when combined with (4.7) and (4.2), implies

$$(\varrho_R U_t^2 - \widehat{E}_{Y_t}) [\partial_X F] = 0, \tag{9.9}$$

where $U_t = \frac{d}{dt} Y_t$. This proves

Theorem 9.1. *The intrinsic velocity U_t of a homentropic acceleration wave satisfies*

$$U_t^2 = \frac{\widehat{E}_{Y_t}}{\varrho_R}, \tag{9.10}$$

where $\widehat{E}_{Y_t}$ is the instantaneous stress-strain modulus at fixed present entropy corresponding to the histories $F^t(Y_t,\cdot)$ and $\eta^t(Y_t,\cdot)$ at the wave.

Theorems 5.1 and 9.1 may be brought together via Remark 8.2 to yield

Remark 9.1. Consider two materials with the same free-energy functional $\mathfrak{p}$ and density ϱ_R, but one, D, a definite conductor and the other, N, a non-conductor. By Theorems 4.2 and 4.3, acceleration waves in D are homothermal with velocity U_t^{D}, while acceleration waves in N are homentropic whenever the corresponding velocity $U_t^{\mathsf{N}} \neq 0$. If there should be two waves, one propagating in N and one propagating in D but both corresponding to the same pair of histories $F^t(Y_t,\cdot)$ and $\vartheta^t(Y_t,\cdot)$, then the velocities of the two waves must obey the following extension of DUHEM's relation*

$$(U_t^{\mathsf{N}})^2 = (U_t^{\mathsf{D}})^2 + \frac{\alpha_{Y_t}^2 \vartheta(Y_t, t)}{\varrho_R^2 \varkappa_{Y_t}}, \tag{9.11}$$

where α_{Y_t} and $\varkappa_{Y_t}$ are the instantaneous stress-temperature modulus and the instantaneous heat capacity at fixed strain, both evaluated at Y_t, t. We notice that since $\vartheta > 0$, the assumption of $\varkappa_{Y_t} > 0$ implies $|U_t^{\mathsf{N}}| \geqq |U_t^{\mathsf{D}}|$, while $\varkappa_{Y_t} > 0$ together with $\alpha_{Y_t} \neq 0$ implies $U_t^{\mathsf{N}} \neq 0$.

10. Growth and Decay of Waves in Non-Conductors

We now seek a formula for the rate of change of the amplitude $a = [\ddot{x}]$ of an acceleration wave propagating with $U_t \neq 0$ into a material which does not conduct heat at all. Since such a wave is homentropic, it follows from (9.3) and Maxwell's theorem that we may use the relations

$$[\ddot{\eta}] = -U[\partial_X \dot{\eta}] = U^2 [\partial_X^2 \eta]. \tag{10.1}$$

To A-1′ and A-2′ we here add another hypothesis.

A-3′**: $\partial_X \eta_r^t(X,\cdot)$ is a smooth function of X and t with respect to the norm $\|\cdot\|_h$ for all X and t.

---

* DUHEM's fundamental relation for elastic materials is explained by TRUESDELL in § 13 of [8].

** In Remark 11.1 we point out that this assumption, which is natural for a homentropic wave, follows from A-1 for a wave entering a homogeneous non-conductor at rest at constant entropy.

It follows from A-1′, A-2′ and our smoothness hypotheses on $\hat{\mathrm{e}}$ that $\partial_X \varepsilon$ and $\dot{\varepsilon}$ are continuous fields in a sufficiently small neighborhood $\mathcal{N}$ of any pair X_0, t_0 with $X_0 \neq Y_{t_0}$. Moreover, from (4.6) with $q \equiv 0$ we get

$$\varrho_R \partial_X \dot{\varepsilon} = T \partial_X \dot{F} + \dot{F} \partial_X T + \varrho_R \partial_X r, \tag{10.2}$$

and hence, using A-1′, A-2′, our smoothness hypotheses for $\widehat{\mathfrak{T}}$, and our assumption of the continuity of $\partial_X r$, we can assert that for a non-conductor $\partial_X \dot{\varepsilon}$ exists and is continuous over $\mathcal{N}$. Therefore

$$\partial_t \partial_X \varepsilon(X, t) = \partial_X \dot{\varepsilon}(X, t), \tag{10.3}$$

and, by (10.2) and (4.5),

$$\varrho_R \partial_t \partial_X \varepsilon(X, t) = T \partial_X \dot{F} + \varrho_R \dot{F} \ddot{x} - \varrho_R \dot{F} b + \varrho_R \partial_X r,$$

except on Σ. Since b, $\partial_X r$ and T are continuous, even across Σ, we have, for non-conductors,

$$[\partial_t \partial_X \varepsilon] = \frac{1}{\varrho_R} T_{Y_t} [\partial_X \dot{F}] + [\ddot{x} \dot{F}] - b_{Y_t} [\dot{F}] \tag{10.4}$$

with $T_{Y_t} = T(Y_t, t)$ and $b_{Y_t} = b(Y_t, t)$.

Clearly, in the present circumstances an equation of the form (6.2) holds for $X \neq Y_t$ with T, $\mathfrak{T}$, ϑ^t, ϑ^t_r, and ϑ replaced by ε, $\hat{\mathrm{e}}$, η^t_r, and η. On applying to that equation the argument giving (6.8), and using, of course, A-1′, A-2′, and A-3′ in place of A-1, A-2, and A-3, we obtain the following expression for $[\partial_t \partial_X \varepsilon]$:

$$\begin{aligned}[\partial_t \partial_X \varepsilon] = {} & \frac{1}{\varrho_R} T_{Y_t} [\partial_X \dot{F}] + \frac{1}{\varrho_R} \widehat{E}_{Y_t} [\dot{F} \partial_X F] + \\ & + \{\mathscr{I}_{Y_t} + \widehat{H}'_{Y_t}(0)\} [\partial_X F] + \vartheta_{Y_t} [\partial_X \dot{\eta}] + \mathscr{J}_{Y_t} [\dot{F}].\end{aligned} \tag{10.5}$$

Here, instead of the coefficients explained in (6.4), (5.3), and (6.6), we have

$$\widehat{E}_{Y_t} = \varrho_R D_F^2 \hat{\mathrm{e}}(F^t(Y_t, \cdot), \eta^t(Y_t, \cdot)) = D_F \widehat{\mathfrak{T}}(F^t(Y_t, \cdot), \eta^t(Y_t, \cdot)), \tag{10.6a}$$

$$\begin{aligned}\mathscr{I}_{Y_t} = {} & \delta_F D_F \hat{\mathrm{e}}(F^t(Y_t, \cdot), \eta^t(Y_t, \cdot) \,|\, \dot{F}^t_r(Y_t, \cdot)) + D_\eta D_F \hat{\mathrm{e}}(F^t(Y_t, \cdot), \eta^t(Y_t, \cdot)) \dot{\eta}_{Y_t} + \\ & + \delta_\eta D_F \hat{\mathrm{e}}(F^t(Y_t, \cdot), \eta^t(Y_t, \cdot) \,|\, \dot{\eta}^t_r(Y_t, \cdot)) \\ = {} & \frac{1}{\varrho_R} \{\delta_F \widehat{\mathfrak{T}}(F^t(Y_t, \cdot), \eta^t(Y_t, \cdot) \,|\, \dot{F}^t_r(Y_t, \cdot)) + \\ & + \hat{\alpha}_{Y_t} \dot{\eta}_{Y_t} + \delta_\eta \widehat{\mathfrak{T}}(F^t(Y_t, \cdot), \eta^t(Y_t, \cdot) \,|\, \dot{\eta}^t_r(Y_t, \cdot))\},\end{aligned} \tag{10.6b}$$

$$\begin{aligned}\mathscr{J}_{Y_t} = {} & D_F \delta_F \hat{\mathrm{e}}(F^t(Y_t, \cdot), \eta^t(Y_t, \cdot) \,|\, \partial_X F^t_r(Y_t, \cdot)) + \\ & + D_F D_\eta \hat{\mathrm{e}}(F^t(Y_t, \cdot), \eta^t(Y_t, \cdot)) (\partial_X \eta)_{Y_t} + D_F \delta_\eta \hat{\mathrm{e}}(F^t(Y_t, \cdot), \vartheta^t(Y_t, \cdot) \,|\, \partial_X \eta^t_r(Y_t, \cdot)) \\ = {} & \frac{1}{\varrho_R} \{\delta_F \widehat{\mathfrak{T}}(F^t(Y_t, \cdot), \eta^t(Y_t, \cdot) \,|\, \partial_X F^t_r(Y_t, \cdot)) + \\ & + \hat{\alpha}_{Y_t} (\partial_X \eta)_{Y_t} + \delta_\eta \widehat{\mathfrak{T}}(F^t(Y_t, \cdot), \eta^t(Y_t, \cdot) \,|\, \partial_X \eta^t_r(Y_t, \cdot))\},\end{aligned} \tag{10.6c}$$

$$\vartheta_{Y_t} = D_\eta \hat{\mathrm{e}}(F^t(Y_t, \cdot), \eta^t(Y_t, \cdot)) = \hat{\mathfrak{t}}(F^t(Y_t, \cdot), \eta^t(Y_t, \cdot)), \tag{10.6d}$$

$$\hat{\alpha}_{Y_t} = D_\eta D_F \hat{\mathrm{e}}(F^t(Y_t, \cdot), \eta^t(Y_t, \cdot)) = \varrho_R D_\eta \widehat{\mathfrak{T}}(F^t(Y_t, \cdot), \eta^t(Y_t, \cdot)), \tag{10.6e}$$

$$\dot{\eta}_{Y_t} = \partial_t \eta(X, t)|_{X = Y_t}, \qquad (\partial_X \eta)_{Y_t} = \partial_X \eta(X, t)|_{X = Y_t}, \tag{10.6f}$$

$$\widehat{H}'_{Y_t}(0) = \frac{\partial}{\partial s} \widehat{\mathfrak{H}}(F^t(Y_t, \cdot), \eta^t(Y_t, \cdot); s)\Big|_{s=0}. \tag{10.6g}$$

In writing the second form of the equations (10.6a)—(10.6e), we have used the stress relation (8.4), the temperature relation (8.5), and, occasionally, the fact that the operators D_F, D_η, δ_F, δ_η commute. The functional $\widehat{\mathfrak{H}}$ of (10.6g) is that defined in (8.25)—(8.28).

Equating (10.4) to (10.5) and using (9.10), (10.1), and the compatibility condition (4.2), we complete the proof of

Theorem 10.1. *For an acceleration wave propagating into a non-conductor*

$$\vartheta_{Y_t}[\ddot{\eta}](t) = \left\{\frac{\mathscr{I}_{Y_t} + \widehat{H}'_{Y_t}(0)}{U_t} - \mathscr{J}_{Y_t} - b_{Y_t}\right\} a(t) + U_t^3[\dot{F}\,\partial_X F] - U_t[\ddot{x}\dot{F}] \tag{10.7}$$

with $\mathscr{I}_{Y_t}$, $\mathscr{J}_{Y_t}$, ϑ_{Y_t}, $\widehat{H}'_{Y_t}(0)$ *given by* (10.6) *and* U_t *by* (9.10).

Suppose $U_t > 0$; then we have both (6.15) and the analogous equation

$$[\ddot{x}\dot{F}] = \ddot{x}_{Y_t^+}[\dot{F}] + \dot{F}_{Y_t^+}[\ddot{x}] + [\ddot{x}][\dot{F}]. \tag{10.8}$$

Combining (10.8) and (4.2), we get

$$[\ddot{x}\dot{F}] = \{\dot{F}_{Y_t^+} - U_t^{-1}\ddot{x}_{Y_t^+}\} a(t) - U_t^{-1} a(t)^2. \tag{10.9}$$

Substitution of this and (6.16) into (10.8) yields us

Remark 10.1. In a non-conductor the kinematical amplitude $a = [\ddot{x}]$ and the "entropic amplitude" $[\ddot{\eta}]$ of an acceleration wave with $U_t > 0$ are related as follows:

$$\vartheta_{Y_t}[\ddot{\eta}](t) = \left\{\frac{\widehat{H}'_t(0) + \mathscr{I}_{Y_t}}{U_t} - \mathscr{J}_{Y_t} - b_{Y_t} + \ddot{x}_{Y_t^+} - U^2(\partial_X F)_{Y_t^+}\right\} a(t). \tag{10.10}$$

With the intention of using (6.1), we now calculate $[\partial_t \partial_X T]$ for a homentropic wave. Replacement of $\mathfrak{T}$, ϑ^t, ϑ^t_r, and ϑ in (6.3) by $\widehat{\mathfrak{T}}$, η^t, η^t_r, and η gives us an expression for $\partial_t \partial_X T$ valid here for $X \neq Y_t$, and it should be clear that if we apply to that expression the argument which previously gave us (6.8), we must now obtain

$$\begin{aligned}[\partial_t \partial_X T] = {} & \widehat{E}_{Y_t}[\partial_X \dot{F}] + \widehat{\widehat{E}}_{Y_t}[\dot{F}\,\partial_X F] + \\ & + \{\hat{I}_{Y_t} + \widehat{G}'_{Y_t}(0)\}[\partial_X F] + \hat{\alpha}_{Y_t}[\partial_X \dot{\eta}] + \hat{J}_{Y_t}[\dot{F}]\end{aligned} \tag{10.11}$$

with the coefficients $\widehat{E}_{Y_t}$, $\hat{\alpha}_{Y_t}$ given by $(10.6\text{a})_2$ and $(10.6\text{e})_2$, instead of (5.3), and with the following equations replacing (6.4a)—(6.4c) and (6.6).

$$\widehat{\widehat{E}}_{Y_t} = D_F^2 \widehat{\mathfrak{T}}(F^t(Y_t,\cdot), \eta^t(Y_t,\cdot)), \tag{10.12a}$$

$$\begin{aligned}\hat{I}_{Y_t} = {} & \delta_F D_F \widehat{\mathfrak{T}}(F^t(Y_t,\cdot), \eta^t(Y_t,\cdot) \mid \dot{F}^t_r(Y_t,\cdot), + D_\eta D_F \widehat{\mathfrak{T}}(F^t(Y_t,\cdot), \eta^t(Y_t,\cdot))\,\dot{\eta}_{Y_t} + \\ & + \delta_\eta D_F \widehat{\mathfrak{T}}(F^t(Y_t,\cdot), \eta^t(Y_t,\cdot) \mid \dot{\eta}^t_r(Y_t,\cdot)),\end{aligned} \tag{10.12b}$$

$$\begin{aligned}\hat{J}_{Y_t} = {} & D_F \delta_F \mathfrak{T}(F^t(Y_t,\cdot), \eta^t(Y_t,\cdot) \mid \partial_X F^t_r(Y_t,\cdot)) + \\ & + D_F \delta_\eta \widehat{\mathfrak{T}}(F^t(Y_t,\cdot), \eta^t(Y_t,\cdot) \mid \partial_X \eta^t_t(Y_t,\cdot)) + \\ & + D_F D_\eta \widehat{\mathfrak{T}}(F^t(Y_t,\cdot), \eta^t(Y_t,\cdot))\,(\partial_X \eta)_{Y_t},\end{aligned} \tag{10.12c}$$

$$\widehat{G}'_{Y_t}(0) = \frac{\partial}{\partial s}\widehat{\mathfrak{G}}(F^t(Y_t,\cdot), \eta^t(Y_t,\cdot); s)\Big|_{s=0}, \tag{10.12d}$$

$$\widehat{E}_{Y_t} = \widehat{G}_{Y_t}(0) = \widehat{\mathfrak{G}}(F^t(Y_t,\cdot), \eta^t(Y_t,\cdot); 0). \tag{10.12e}$$

We may note that it follows from the stress relation (8.4) and the equations (10.12) that $\hat{\hat{E}}_{Y_t}$, $\hat{I}_{Y_t}$, $\hat{J}_{Y_t}$, $\hat{G}'_{Y_t}(0)$, and $\hat{E}_{Y_t}$ are determined when $F^t(Y_t,\cdot)$, $\eta^t(Y_t,\cdot)$, $(\partial_X\eta)_{Y_t}$ and either the functional $\hat{\mathfrak{T}}$ or the functional $\hat{\mathfrak{e}}$ are given. From the equations (9.10), (4.2), (10.1), (10.7), and (10.11) we get

$$\begin{aligned}\frac{1}{\varrho_R}[\partial_t\partial_X T] = U_t^2[\partial_X\dot{F}] + \frac{1}{\varrho_R}\left\{\hat{\hat{E}}_{Y_t} - \frac{\hat{\alpha}_{Y_t}U_t^2}{\vartheta_{Y_t}}\right\}[\dot{F}\,\partial_X F] + \frac{\hat{\alpha}_{Y_t}}{\varrho_R\vartheta_{Y_t}}[\ddot{x}\dot{F}] + \\ + \frac{1}{\varrho_R U_t}\left\{\frac{\hat{I}_{Y_t}+\hat{G}'_{Y_t}(0)}{U_t} - \hat{J}_{Y_t} - \frac{\hat{\alpha}_{Y_t}}{\vartheta_{Y_t}}\left(\frac{\mathscr{I}_{Y_t}+\hat{H}'_{Y_t}(0)}{U_t} - \mathscr{J}_{Y_t} - b_{Y_t}\right)\right\}a(t),\end{aligned} \tag{10.13}$$

and substitution of this result into (6.1) yields

Theorem 10.2. *In a non-conductor the amplitude $a=a(t)$ of an acceleration wave obeys the relation*

$$\frac{da}{dt} = -\hat{\mu}_t a + (\hat{\beta}_t - U_t^2\hat{\gamma}_t)[\dot{F}\,\partial_X F] + \hat{\gamma}_t[\ddot{x}\dot{F}], \tag{10.14}$$

where

$$\hat{\mu}_t = -\frac{1}{2U_t}\frac{dU_t}{dt} + \frac{1}{2\varrho_R U_t}\left\{\hat{J}_{Y_t} - \frac{\hat{I}_{Y_t}+\hat{G}'_{Y_t}(0)}{U_t} - \frac{\hat{\alpha}_{Y_t}}{\vartheta_{Y_t}}\left(\frac{\mathscr{I}_{Y_t}+\hat{H}'_{Y_t}}{U_t} - \mathscr{J}_{Y_t} - b_{Y_t}\right)\right\}, \tag{10.15}$$

$$\hat{\beta}_t = \frac{\hat{\hat{E}}_{Y_t}}{2\varrho_R}, \tag{10.16}$$

and

$$\hat{\gamma}_t = \frac{\hat{\alpha}_{Y_t}}{2\varrho_R\vartheta_{Y_t}}; \tag{10.17}$$

U_t is the intrinsic velocity of the wave; the numbers $\hat{\hat{E}}_{Y_t}$, $\mathscr{I}_{Y_t}$, $\mathscr{J}_{Y_t}$, $\hat{\vartheta}_{Y_t}$, $\hat{\alpha}_{Y_t}$, $\hat{H}'_{Y_t}(0)$, $\hat{I}_{Y_t}$, $\hat{J}_{Y_t}$, and $\hat{G}'_{Y_t}(0)$, given by equations (10.6) and (10.12), are determined when $F^t(Y_t,\cdot)$, $\eta^t(Y_t,\cdot)$, $(\partial_X\eta)_{Y_t}$ and the functional $\hat{\mathfrak{e}}$ are specified.

Substituting (6.16) and (10.9) into (10.14), we obtain the following analogue of Remark 6.1.

Remark 10.2. The amplitude of an acceleration wave with $U_t>0$ propagating in a non-conductor obeys the differential equation

$$\frac{da}{dt} = \{\hat{\beta}_{Y_t}U_t^{-2}\dot{F}_{Y_t^+} - (\hat{\beta}_t - U_t^2\hat{\gamma}_t)\,U_t^{-1}(\partial_X F)_{Y_t^+} - \hat{\gamma}_t U_t^{-1}\ddot{x}_{Y_t^+} - \hat{\mu}_t\}\,a - U_t^{-3}\hat{\beta}_t a^2 \tag{10.18}$$

with $\hat{\mu}_t$, $\hat{\beta}_t$, and $\hat{\gamma}_t$ given by (10.15)—(10.17).

11. A Wave Entering a Non-Conductor at Rest in a Homogeneous State

We here consider an acceleration wave which since time $t=0$ has been propagating into a non-conductor which previously had been at rest in a homogeneous configuration at constant and uniform entropy; thus, we suppose that $U_t>0$ and that for $X>Y_t$ and $t\geqq 0$

$$x = x(X, t-s) = F_0 X + X_0, \qquad 0\leqq s<\infty, \tag{11.1a}$$

$$\eta = \eta(X, t-s) = \eta_0, \qquad 0\leqq s\leqq\infty. \tag{11.1b}$$

It follows from (8.1c) that the temperature ahead of such a wave is also constant and uniform; *i.e.*, (11.1) implies (7.1). (Actually, the assumption that the functional transformation (8.2) is one-to-one renders the hypothesis (7.1) *equivalent* to (11.1).) Before giving the solution of (10.14) corresponding to (11.1) we state, without giving the proof, the following obvious analogue for non-conductors of Remark 7.1.

Remark 11.1. If the material is a non-conductor, and if (11.1) holds with $U_t > 0$, then the hypothesis A-1′ implies A-2′ and A-3′ for $X > Y_0$ and $t > 0$.

Letting $F_o^\dagger$ and $\eta_o^\dagger$ be the constant histories defined by

$$F_o^\dagger(s) = F_o, \qquad \eta_o^\dagger(s) = \eta_o, \qquad 0 \leqq s < \infty, \tag{11.2}$$

we observe that, by (11.1), for $t \geqq 0$ we here have

$$F^t(Y_t, \cdot) = F_o^\dagger, \qquad \vartheta^t(Y_t, \cdot) = \vartheta_o^\dagger, \tag{11.3}$$

$$\ddot{x}(Y_t^+, t) = \dot{F}(Y_t^+, t) = \partial_X F^t(Y_t^+, s) = \dot{\eta}^t(Y_t^+, s) = \partial_X \eta^t(Y_t^+, s) = 0, \quad 0 \leqq s < \infty. \tag{11.4}$$

Since a wave in a non-conductor obeys (9.3), the last two equalities in (11.4) yield

$$\dot{\eta}_{Y_t} = (\partial_X \eta)_{Y_t} = 0, \qquad t \geqq 0, \tag{11.5}$$

for the quantities defined in (10.6f). Furthermore, since, by A-2′ and A-3′, $\dot{F}_r^t$, $\partial_X F_r^t$, $\dot{\eta}_r^t$, and $\partial_X \eta_r^t$, as points $\mathscr{H}_h$, are continuous functions of X at Y_t, it also follows from (11.4) that, for $t \geqq 0$, $\dot{F}_r^t(Y_t, \cdot)$, $\partial_X F_r^t(Y_t, \cdot)$, $\dot{\eta}_r^t(Y_t, \cdot)$, and $\partial_X \eta_r^t(Y_t, \cdot)$ are all constant functions with value zero. These observations, (10.12), (10.6), and (9.10), make it apparent that here for $t \geqq 0$, $\hat{G}_{Y_t}(0)$, $\hat{G}'_{Y_t}(0)$, $\hat{\hat{E}}_{Y_t}$, and U_t are constants,

$$\hat{G}_{Y_t}(0) = \hat{E}_{Y_t} = \hat{G}(0), \qquad \hat{G}'_{Y_t}(0) = \hat{G}'(0), \qquad \hat{\hat{E}}_{Y_t} = \hat{\hat{E}}, \qquad U_t = U, \tag{11.6}$$

given by

$$\begin{aligned} \hat{G}(0) &= \hat{\mathfrak{G}}(F_o^\dagger, \eta_o^\dagger; 0) = D_F \hat{\mathfrak{T}}(F_o^\dagger, \eta_o^\dagger), \\ \hat{G}'(0) &= \frac{\partial}{\partial s} \hat{\mathfrak{G}}(F_o^\dagger, \eta_o^\dagger; s)\Big|_{s=0}, \\ \hat{\hat{E}} &= D_F^2 \hat{\mathfrak{T}}(F_o^\dagger, \eta_o^\dagger), \end{aligned} \tag{11.7}$$

$$U = \sqrt{\frac{\hat{G}(0)}{\varrho_R}}. \tag{11.8}$$

For the other quantities in (10.6) and (10.12) we clearly have

$$\hat{I}_{Y_t} = \hat{J}_{Y_t} = \mathscr{I}_{Y_t} = \mathscr{J}_{Y_t} = \hat{H}'_{Y_t}(0) = b_{Y_t} = 0, \qquad \hat{\alpha}_{Y_t} = \hat{\alpha}, \qquad \vartheta_{Y_t} = \vartheta_o, \tag{11.9}$$

where

$$\hat{\alpha} = D_\eta \hat{\mathfrak{T}}(F_o^\dagger, \eta_o^\dagger) \quad \text{and} \quad \vartheta_o = \hat{\mathfrak{t}}(F_o^\dagger, \eta_o^\dagger) \tag{11.10}$$

are constants; in writing $\hat{H}'_{Y_t}(0) = 0$ we make use of Remark 8.5, a consequence of the internal dissipation inequality. The assertion that $b_{Y_t} = 0$ follows immediately from (11.1), (4.5), (8.1b), and our smoothness hypothesis on the field b. Consequences of (11.4) and (4.2) are

$$\begin{aligned} [\dot{F}\,\partial_X F] &= [\partial_X F]\,[\dot{F}] = -U^{-3}a^2, \\ [\ddot{x}\dot{F}] &= [\ddot{x}]\,[\dot{F}] = -U^{-1}a^2. \end{aligned} \tag{11.11}$$

It follows from (10.15)—(10.16), (11.6), (11.9) and (11.11) that whenever (11.1) holds, (10.14) takes the form

$$\frac{da}{dt} = -\hat{\mu}\,a + \frac{\hat{\mu}}{\hat{\lambda}}\,a^2, \tag{11.12}$$

where

$$\hat{\mu} = -\frac{\widehat{G}'(0)}{2\widehat{G}(0)} = \text{const.}, \qquad \hat{\lambda} = \frac{\widehat{G}'(0)\,U}{\hat{\hat{E}}} = \text{const.} \tag{11.13}$$

Hence we have

Theorem 11.1. *Consider an acceleration wave propagating, since $t=0$, into a non-conductor of infinite extent which previous to arrival of the wave had always been at rest with its deformation gradient F_0 and entropy η_0 constant in space and time. Then, for $t\geqq 0$ the intrinsic velocity is a constant given by* (11.8), *and the amplitude is the following explicit function of t:*

$$a(t) = \frac{\hat{\lambda}}{\left(\dfrac{\hat{\lambda}}{a(0)} - 1\right)e^{\hat{\mu} t} - 1}. \tag{11.14}$$

For a given non-conductor, $\hat{\mu}$ and $\hat{\lambda}$ are constants depending on only F_0 and η_0 through the functional $\hat{\mathfrak{T}}$ as shown in (11.13), (11.7), *and* (11.8).

Setting the above theorem next to Theorem 4.1 of II, we make the following observation.

Remark 11.2. If the material does not conduct heat, the equations (4.5), (4.11), and (4.12) of II, giving the intrinsic velocity and amplitude of an acceleration wave entering region at rest in a uniform state, are correct in our present more general theory, provided that the relaxation function and second-order modulus used in II are taken at fixed entropy.

This remark provides a transformation into our present context of the conclusions of Sections 4—6 of II. Unfortunately, such a simple rule for supressing thermodynamic variables generally ceases to remain valid when the region ahead of the wave is not in equilibrium. On comparing our present Theorem 10.2 with Theorem 2.1 of II we see that *not all the formulae of* II *hold for a non-conductor with the D_F and δ_F derivatives taken at fixed entropy.*

It follows from (11.4) and (11.5) that whenever (11.1) is assumed equation (10.10) reduces to $\vartheta_{Y_t}[\ddot{\eta}](t)=\{0\}\,a(t)$, and hence, by (10.1), we have

Remark 11.3. If the hypothesis of Theorem 11.1 is assumed, then for $t\geqq 0$ we have

$$[\ddot{\eta}] = [\partial_X\,\dot{\eta}] = [\partial_X^2\eta] = 0 \tag{11.15}$$

regardless of the magnitude of $[\ddot{x}]$.

We acknowledge with gratitude the support of this research by the National Science Foundation, the Air Force Office of Scientific Research, and the Advanced Research Projects Agency of the Department of Defense.

References

[1] COLEMAN, B. D., M. E. GURTIN, & I. HERRERA R., Arch. Rational Mech. Anal. **19**, 1—18 (1965).
[2] COLEMAN, B. D., & M. E. GURTIN, Arch. Rational Mech. Anal. **19**, 239—265 (1965).
[3] COLEMAN, B. D., Arch. Rational Mech. Anal. **17**, 1—46 (1964).
[4] COLEMAN, B. D., Arch. Rational Mech. Anal. **17**, 230—254 (1964).
[5] TRUESDELL, C., & R. A. TOUPIN, The Classical Field Theories. In the Encyclopedia of Physics, Vol. III/1, edited by S. FLÜGGE. Berlin-Göttingen-Heidelberg: Springer 1960.
[6] FOURIER, J., Théorie Analytique de la Chaleur. Paris, 1822 = Oeuvres, Vol. 1.
[7] FLORY, P. J., Principles of Polymer Chemistry. Ithaca, New York: Cornell University Press 1953.
[8] TRUESDELL, C., Arch. Rational Mech. Anal. **8**, 263—296 (1961).
[9] VARLEY, E., Arch. Rational Mech. Anal. **19**, 215—225 (1965).
[10] COLEMAN, B. D., & W. NOLL, Arch. Rational Mech. Anal. **13**, 167—178 (1963).
[11] RANKINE, W. J. M., Phil. Trans. Roy. Soc. London **160**, 277—288 (1870).
[12] HAYES, W. D., Gasdynamic Discontinuities. Princeton: Princeton University Press 1960.

Mellon Institute
Pittsburgh
and
Brown University
Providence

(Received March 17, 1965)

IV. Thermodynamics and the Velocity of General Acceleration Waves

BERNARD D. COLEMAN & MORTON E. GURTIN

Contents

Preface

Here we shall discuss the propagation of three-dimensional acceleration waves in materials with memory.*

To fix ideas, let us recall some results from the theory of finite elasticity. If we express the position $\boldsymbol{x}$ of a material point X at time t as a function of t and the position $\boldsymbol{X}$ of X in some fixed reference configuration $\mathscr{R}$, then the tensor $\boldsymbol{F}$ with Cartesian components

$$F_{ij} = \frac{\partial}{\partial X_j} x_i(\boldsymbol{X}, t) \tag{1}$$

is called the deformation gradient for X at time t.**

We may take the *Piola-Kirchhoff stress tensor* $\boldsymbol{S}$ to be defined by the formula***

$$\boldsymbol{S} = \frac{1}{\varrho} \boldsymbol{T}(\boldsymbol{F}^T)^{-1} \tag{2}$$

* References [*1*], [*2*], and [*3*] are the papers "I", "II", and "III" in this series; they deal with one-dimensional motions.

** We denote vectors, tensors, and points in Euclidean space by boldface letters. We assume throughout that a single, fixed, Cartesian coordinate system is given. If $\boldsymbol{n}$ is a vector and $\boldsymbol{X}$ a point, then n_i and X_j denote their Cartesian components. If $\boldsymbol{A}$ is a tensor, then $\boldsymbol{A}^T$ denotes its transpose, $\boldsymbol{A}^{-1}$ its inverse, A_{ij} its Cartesian components, and $|\boldsymbol{A}| = \sqrt{A_{ij}A_{ij}}$ its "absolute value". Summation over repeated subscripts is understood.

*** The tensor $\boldsymbol{S}$ defined in (1.2) differs by but a constant factor ϱ_R, the mass density in the reference configuration, from the tensor which TRUESDELL [*4*] calls the *first Piola-Kirchhoff tensor*.

where $\boldsymbol{T}$ is the familiar stress tensor of Cauchy and ϱ is the present mass density. In elasticity theory, when thermodynamic influences are ignored, $\boldsymbol{S}$ is a function of $\boldsymbol{F}$:

$$\boldsymbol{S}=\boldsymbol{S}(\boldsymbol{F}). \tag{3}$$

The classical **Fresnel-Hadamard theorem*** asserts that for every material obeying (3) the amplitude** $\boldsymbol{s}$ and the speed of propagation V of an acceleration wave traveling in the direction $\boldsymbol{n}$ must obey the *propagation condition*

$$\boldsymbol{Q}(\boldsymbol{n})\,\boldsymbol{s}=V^2\boldsymbol{s}, \tag{4}$$

where the tensor $\boldsymbol{Q}(\boldsymbol{n})$, called the *acoustic tensor*, has the Cartesian components

$$Q_{ij}(\boldsymbol{n})=\frac{\partial S_{il}(\boldsymbol{F})}{\partial F_{jm}}F_{pm}F_{ql}n_p n_q. \tag{5}$$

A second result in the theory of elastic wave-propagation is the **Ericksen-Truesdell theorem*****: All higher-order discontinuities also obey the Fresnel-Hadamard propagation condition (4).

Of course, even in elasticity theory one should include thermodynamic influences and allow $\boldsymbol{S}$ to depend not only on $\boldsymbol{F}$ but also on a thermodynamic variable. When this is done, the **first theorem of Duhem**† tells us that homothermal (homentropic) acceleration waves in elastic materials obey (4), provided the derivative $\frac{\partial}{\partial F_{jm}}$ in (5) is taken at constant temperature (entropy). The physical circumstances in which acceleration waves are homothermal or homentropic are given in the **second theorem of Duhem**: Every acceleration wave in an elastic material obeying Fourier's law of heat conduction with positive-definite conductivity is homothermal; every acceleration wave in an elastic non-conductor of heat is homentropic.

In this paper we extend each of the foregoing theorems to general materials with fading memory. We show that the theory of acceleration waves in such materials can be based on the notion of an *instantaneous acoustic tensor*.†† This tensor is defined by a formula analogous to (5); the partial derivatives shown

---

* For the most general statement of the Fresnel-Hadamard theorem, and for a detailed discussion of its consequences, see TRUESDELL [*4*]; he appears to be the first to observe that this theorem remains valid even when it is *not* assumed that the stress-strain relation is derived from a stored energy function.

** We use the terminology of TRUESDELL & TOUPIN [*5*] whenever it does not conflict with that of other papers in this series. The *speed of propagation* is the normal speed of a singular surface with respect to the particles instantaneously situated upon it. The analogue of V in our one-dimensional theory is *not* the *intrinsic velocity* U of papers I, II, and III, but rather the *relative velocity* defined in eq. (5.8) of I. Though we can use the terminology, we cannot imitate the notation of [*5*], for in that article the speed of propagation is denoted by U. In II and III we took the jump in the acceleration as a measure of the "amplitude" of an acceleration wave. It follows from eq. $(2.11)_1$ that $[\ddot{\boldsymbol{x}}]=V^2\boldsymbol{s}$. Our present $\boldsymbol{s}$ is written $\boldsymbol{s}_0$ in § 190 of [*5*].

*** This assertion, known to hold for perfect fluids, was established for isotropic incompressible hyperelastic materials by ERICKSEN [*6*] and for compressible elastic materials by TRUESDELL [*4*].

† The two theorems of DUHEM are explained and given modern proofs by TRUESDELL [*4*].

†† This should not be surprising in view of the results of I and III.

there are, however, replaced by "instantaneous partial derivatives" $D_{F_{jm}}$ of the stress functional.

Among our results is a proof of the assertion that, even in a material with memory, all waves of order equal to or greater than two must obey the Fresnel-Hadamard propagation condition with the same instantaneous acoustic tensor. For homothermal (homentropic) waves the derivatives $D_{F_{jm}}$ are to be taken with the history of the temperature (entropy) held fixed. If we define a *definite conductor* to be a material for which the gradient of the heat flux with respect to the temperature gradient is always a negative-definite tensor, then every acceleration wave in a definite conductor with memory is homothermal. The "instantaneous derivatives" occurring in our present theory are precisely the "D-operators" appearing in the generalized-stress relation which COLEMAN [*7, 8*] has derived in his treatment of the thermodynamics of materials with memory. We here use the generalized-stress relation to show that every acceleration wave in a non-conductor of heat is homentropic.

A well known **theorem of Hadamard*** asserts that for an elastic material with a stored energy function the acoustic tensor (5) must be symmetric:

$$Q_{ij}(\boldsymbol{n}) = Q_{ji}(\boldsymbol{n}). \tag{6}$$

It follows from this theorem, the first theorem of DUHEM, and familiar relations of classical thermodynamics, that the acoustic tensor of an elastic material must be symmetric in homothermal and homentropic waves. Using the generalized stress-relation, we here show that the symmetry relation (6) must hold for *every* material with fading memory which is either a definite conductor or a non-conductor of heat. This, our main result, should enable an experimenter to test the physical appropriateness of the generalized-stress relation.**

1. Mathematical Preliminaries

As in the previous papers in this series, we here use the expression "continuously Fréchet-differentiable" in the following precise sense.

Let $\mathscr{A}$ and $\mathscr{B}$ denote normed linear spaces, with norms $\|\cdot\|_{\mathscr{A}}$, $\|\cdot\|_{\mathscr{B}}$, and let $\mathscr{N}$ be an open subset of $\mathscr{A}$. We say that a mapping $a \to f(a)$ of $\mathscr{N}$ into $\mathscr{B}$ is *N-times continuously Fréchet-differentiable* if the following two conditions hold:

(i) for each n, $n = 1, 2, \ldots, N$, there exists a function

$$(a, a_1, a_2, \ldots, a_n) \to \delta^n f(a \mid a_1, a_2, \ldots, a_n)$$

mapping $\mathscr{N} \times \mathscr{A} \times \mathscr{A} \times \cdots \times \mathscr{A}$ into $\mathscr{B}$ that is multilinear and symmetric in the n variables $a_1, a_2, \ldots, a_n$ and jointly continuous in the $n+1$ variables $a, a_1, a_2, \ldots, a_n$;

(ii) for each a in $\mathscr{N}$

$$\left\| f(a+\omega) - f(a) - \sum_{n=1}^{N} \frac{1}{n!}\, \delta^n f(a \mid \omega, \omega, \ldots, \omega) \right\|_{\mathscr{B}} = o\left(\|\omega\|_{\mathscr{A}}^{N}\right). \tag{1.1}$$

The function $\delta^n f$ is the nth *Fréchet-derivative* of f.

* [*9*]; see also [*4*], § 12, where it is pointed out that the theorem of HADAMARD generalizes earlier results of CHRISTOFFEL and HUGONIOT.

** Fortunately, TRUESDELL [*4, 10*] has derived experimentally verifiable relations which depend on the symmetry of the acoustic tensor.

Henceforth $\mathfrak{H}$ ($\mathscr{H}$) will denote a normed linear space, with norm $\|\cdot\|$, whose elements are tensor-valued (real-valued) functions on $(0, \infty)$.

We say that a tensor-valued (real-valued) function on $[0, \infty)$ belongs to $\mathfrak{H}^{\oplus}$ ($\mathscr{H}^{\oplus}$) if its restriction to $(0, \infty)$ belongs to $\mathfrak{H}$ ($\mathscr{H}$). We define the norm $\|f\|^{\oplus}$ of a function f in $\mathfrak{H}^{\oplus}$, or $\mathscr{H}^{\oplus}$, to be

$$\|f\|^{\oplus}=|f(0)|+\|f_r\|, \tag{1.2}$$

where f_r is the restriction of f to $(0, \infty)$. In thermodynamics we frequently use the product space $\mathscr{S}=\mathfrak{H}^{\oplus}\times\mathscr{H}^{\oplus}$ with the "direct sum norm"; *i.e.*, if $\boldsymbol{\Lambda}=(\boldsymbol{G}, g)$ is in $\mathscr{S}$, then

$$\|\boldsymbol{\Lambda}\|^{\oplus}=\|\boldsymbol{G}\|^{\oplus}+\|g\|^{\oplus}. \tag{1.3}$$

The use of the one symbol $\|\cdot\|^{\oplus}$ for the norms on $\mathfrak{H}^{\oplus}$, $\mathscr{H}^{\oplus}$, and $\mathfrak{H}^{\oplus}\times\mathscr{H}^{\oplus}$ should not be troublesome; there will never be any doubt as to which of these three spaces contains a given function.

Some of our theorems are sufficiently general that the norm $\|\cdot\|$ on $\mathfrak{H}$ or $\mathscr{H}$ need not be specified. In certain applications, however, $\mathfrak{H}$ and $\mathscr{H}$ will be Hilbert spaces of the type which occur in the theory of fading memory [*11*]. In that theory one assumes that a fixed, positive, continuous, monotone-decreasing, square integrable function h is given over $(0, \infty)$. This *influence function* is used to compute $\|f_r\|$ as follows:

$$\|f_r\|=\sqrt{\int_0^{\infty}|f_r(s)|^2 h(s)^2\, ds}, \tag{1.4}$$

and $\mathfrak{H}$ (or $\mathscr{H}$) is then the space of functions f_r (regarding as equivalent those functions which differ on only sets of measure zero) for which $\|f_r\|$ is finite.

2. Propagation in the Absence of Thermodynamic Influences

Let the function $\boldsymbol{F}^t$ on $[0, \infty)$ be the *history up to time t of the deformation gradient* at a particular material point $\boldsymbol{X}$:

$$\boldsymbol{F}^t(s)=\boldsymbol{F}(\boldsymbol{X}, t-s), \qquad 0\leqq s<\infty. \tag{2.1}$$

In this section we assume that the Piola-Kirchhoff tensor $\boldsymbol{S}$ is given by a functional $\mathfrak{S}$ of $\boldsymbol{F}^t$:

$$\boldsymbol{S}(t)=\mathfrak{S}(\boldsymbol{F}^t). \tag{2.2}$$

In writing (2.2) we are ignoring thermodynamic influences; this apparent neglect will be rectified in Sections 3 and 6.

In order to discuss waves, we need a *smoothness hypothesis on the constitutive functional* $\mathfrak{S}$. We assume that $\mathfrak{S}$ has for its domain of definition an open subset $\mathscr{D}$ of a normed function space $\mathfrak{H}^{\oplus}$ and is continuously Fréchet-differentiable throughout this domain. Of course, the elements of $\mathfrak{H}^{\oplus}$ are tensor-valued functions $\boldsymbol{F}^t$ on $[0, \infty)$. The norm $\|\cdot\|^{\oplus}$ over $\mathfrak{H}^{\oplus}$ is given by (1.2):

$$\|\boldsymbol{F}^t\|^{\oplus}=|\boldsymbol{F}(t)|+\|\boldsymbol{F}_r^t\|; \tag{2.3}$$

$\boldsymbol{F}(t)=\boldsymbol{F}^t(0)$ is called the *present value*, and $\boldsymbol{F}_r^t$, the restriction of $\boldsymbol{F}^t$ to $(0, \infty)$, is called the *past history* of the deformation gradient. The space of past histories is denoted by $\mathfrak{H}$. The generality of our present discussion is such that we need not specify the norm $\|\cdot\|$ in (2.3). This norm over $\mathfrak{H}$ may, however, be thought of as the norm (1.4) of the theory of fading memory.

For convenience we assume that the material under consideration is *homogeneous* and that a homogeneous configuration $\mathscr{R}$ has been taken as reference; thus, although $\boldsymbol{F}^t$ and $\boldsymbol{S}(t)$ generally depend on $\boldsymbol{X}$ as well as t, the functional $\mathfrak{S}$ and the mass density ϱ_R in $\mathscr{R}$ are independent of $\boldsymbol{X}$.*

Since a knowledge of the function $\boldsymbol{F}^t$ is equivalent to a knowledge of the function $\boldsymbol{F}_r^t$ and the tensor $\boldsymbol{F}(t)$, we can write the constitutive equation (2.2) in the form

$$\boldsymbol{S}=\mathfrak{S}(\boldsymbol{F}_r^t;\boldsymbol{F}). \tag{2.4**}$$

We here use the abbreviations $\boldsymbol{S}=\boldsymbol{S}(t)$, $\boldsymbol{F}=\boldsymbol{F}(t)$.

The assumed smoothness of the functional $\mathfrak{S}$ guarantees the existence of the following partial derivatives for every function $\boldsymbol{F}^t$ in $\mathscr{D}$, every function $\boldsymbol{G}_r$ in $\mathscr{H}$, and every tensor $\boldsymbol{A}$:

$$D_{\boldsymbol{F}}\mathfrak{S}(\boldsymbol{F}^t|\boldsymbol{A})\overset{\text{def}}{=}\frac{\partial}{\partial\lambda}\mathfrak{S}(\boldsymbol{F}_r^t;\boldsymbol{F}+\lambda\boldsymbol{A})\Big|_{\lambda=0}, \tag{2.5a}$$

$$\delta_{\boldsymbol{F}_r^t}\mathfrak{S}(\boldsymbol{F}^t|\boldsymbol{G}_r)\overset{\text{def}}{=}\frac{\partial}{\partial\lambda}\mathfrak{S}(\boldsymbol{F}_r^t+\lambda\boldsymbol{G}_r;\boldsymbol{F})\Big|_{\lambda=0}. \tag{2.5b}$$

Furthermore, $D_{\boldsymbol{F}}\mathfrak{S}(\boldsymbol{F}^t|\boldsymbol{A})$ is linear in $\boldsymbol{A}$ and jointly continuous in $\boldsymbol{F}^t$ and $\boldsymbol{A}$; $\delta_{\boldsymbol{F}_r^t}\mathfrak{S}(\boldsymbol{F}^t|\boldsymbol{G}_r)$ is linear in $\boldsymbol{G}_r$ and jointly continuous in $\boldsymbol{F}^t$ and $\boldsymbol{G}_r$. In fact, these functions, being, respectively, the Fréchet-derivatives of the mappings $\boldsymbol{F}\to\mathfrak{S}(\boldsymbol{F}_r^t;\boldsymbol{F})$ and $\boldsymbol{F}_r^t\to\mathfrak{S}(\boldsymbol{F}_r^t;\boldsymbol{F})$, are related to the Fréchet-derivative $\delta\mathfrak{S}(\boldsymbol{F}^t|\boldsymbol{G})$ of the mapping $\boldsymbol{F}^t\to\mathfrak{S}(\boldsymbol{F}^t)$ by the equation

$$\delta\mathfrak{S}(\boldsymbol{F}^t|\boldsymbol{G})=D_{\boldsymbol{F}}\mathfrak{S}\big(\boldsymbol{F}^t|\boldsymbol{G}(0)\big)+\delta_{\boldsymbol{F}_r^t}\mathfrak{S}(\boldsymbol{F}^t|\boldsymbol{G}_r), \tag{2.6}$$

which holds for every function $\boldsymbol{G}$ in $\mathscr{H}^{\oplus}$.

Let $D_{\boldsymbol{F}}\mathfrak{S}(\boldsymbol{F}^t)$ denote the fourth-order tensor, with Cartesian components $D_{F_{ij}}\mathfrak{S}_{pq}(\boldsymbol{F}^t)$, characterizing the linear transformation mapping the second-order tensor $\boldsymbol{A}$ into the second-order tensor $D_{\boldsymbol{F}}\mathfrak{S}(\boldsymbol{F}^t|\boldsymbol{A})$ with components $[D_{\boldsymbol{F}}\mathfrak{S}(\boldsymbol{F}^t|\boldsymbol{A})]_{pq}$; *i.e.*,

$$D_{F_{ij}}\mathfrak{S}_{pq}(\boldsymbol{F}^t)A_{ij}=[D_{\boldsymbol{F}}\mathfrak{S}(\boldsymbol{F}^t|\boldsymbol{A})]_{pq}. \tag{2.7}$$

It follows from (2.5a) that

$$D_{F_{ij}}\mathfrak{S}_{pq}(\boldsymbol{F}^t)=\frac{\partial}{\partial F_{ij}}\mathfrak{S}_{pq}(\boldsymbol{F}_r^t;\boldsymbol{F}); \tag{2.8}$$

i.e., $D_{F_{ij}}\mathfrak{S}_{pq}$ is the partial derivative of $\mathfrak{S}_{pq}(\boldsymbol{F}_r^t;\boldsymbol{F})$ with respect to the component F_{ij} of the present value $\boldsymbol{F}$. Since this partial derivative is computed holding the past history $\boldsymbol{F}_r^t$ fixed, we call $D_{F_{ij}}\mathfrak{S}_{pq}(\boldsymbol{F}^t)$ the *instantaneous partial derivative* of $\mathfrak{S}_{pq}(\boldsymbol{F}^t)$ with respect to F_{ij}.

Given any unit vector $\boldsymbol{n}$ and any history $\boldsymbol{F}^t$ in $\mathscr{D}$, we call the tensor $\boldsymbol{Q}(\boldsymbol{n})$ with the components

$$Q_{ij}(\boldsymbol{n})=D_{F_{jm}}\mathfrak{S}_{il}(\boldsymbol{F}^t)F_{pm}F_{ql}n_p n_q \tag{2.9}$$

the *instantaneous acoustic tensor* corresponding to $\boldsymbol{F}^t$ and $\boldsymbol{n}$.

---

* Our results are valid without change for inhomogeneous bodies provided that the dependence of ϱ_R and $\mathfrak{S}$ on $\boldsymbol{X}$ is assumed to be sufficiently smooth.

** No confusion should arise from our use of the same symbol $\mathfrak{S}$ for the functionals in (2.2) and (2.4).

Of course, equation (2.9), which is meaningful for materials with memory, has a formal similarity to equation (5), which is meaningful for only elastic materials. We deliberately use a notation which emphasizes this similarity. It is now our intention to prove that the tensor $\boldsymbol{Q}(\boldsymbol{n})$ defined by (2.9) is a "true acoustic tensor", *i.e.*, that it really enters an equation of the type (4) for acceleration waves.

Acceleration Waves

It is sometimes helpful to make explicit the dependence of $\boldsymbol{F}^t$ and $\boldsymbol{F}^t_r$ on $\boldsymbol{X}$; we do this by writing $\boldsymbol{F}^t(\boldsymbol{X},\cdot)$ and $\boldsymbol{F}^t_r(\boldsymbol{X},\cdot)$ for these functions.

We call a singular surface* of the motion $\boldsymbol{x}(\boldsymbol{X},t)$ an *acceleration wave* if the following two conditions hold.

A-1: $\boldsymbol{x}$, $\dot{\boldsymbol{x}}$, and $\boldsymbol{F}$ are continuous functions of $\boldsymbol{X}$ and t jointly for all $\boldsymbol{X}$ and t, while $\ddot{\boldsymbol{x}}$, $\dot{\boldsymbol{F}}$, and $\nabla_{\boldsymbol{X}}\boldsymbol{F}$ $\left(\text{with components } \frac{\partial}{\partial X_k}F_{ij}\right)$ all suffer jump discontinuities across the surface but are continuous in $\boldsymbol{X}$ and t jointly everywhere else.

A-2: At each time t, the mapping $\boldsymbol{X}\to\boldsymbol{F}^t_r(\boldsymbol{X},\cdot)$, with values in the function space $\mathcal{H}$, is (once) continuously Fréchet-differentiable for all $\boldsymbol{X}$.

We of course also assume that for each $\boldsymbol{X}$ and t the history $\boldsymbol{F}^t(\boldsymbol{X},\cdot)$ belongs to the domain $\mathcal{D}$ of $\mathfrak{S}$. The specific extrinsic body force $\boldsymbol{b}$ is assumed to be continuous for all $\boldsymbol{X}$ and t; hence, the jump $[b_i]$ across the singular surface in each of the components b_i of $\boldsymbol{b}$ is zero:

$$[b_i]=0. \tag{2.10}$$

The condition A-1 is the requirement that an acceleration wave be a singular surface of order 2 in the classification scheme of DUHEM and HADAMARD [*5*, §187]. The condition A-2 asserts that the past history $\boldsymbol{F}^t_r(\boldsymbol{X},\cdot)$ is a smooth function of $\boldsymbol{X}$ with respect to the norm $\|\cdot\|$ of (2.3) for all $\boldsymbol{X}$ and t; this condition serves to limit the wildness of the *past* variations of the strain. If $\|\cdot\|$ is the h-norm (1.4) and if the wave is entering a material which has always been at rest, then A-2 is implied by A-1.**

The condition A-1 implies that the jumps $[\ddot{x}_i]$ and $\left[\frac{\partial}{\partial X_k}F_{ij}\right]$ in the components of $\ddot{\boldsymbol{x}}$ and $\nabla_{\boldsymbol{X}}\boldsymbol{F}$ must obey the following kinematical *conditions of compatibility*:***

$$[\ddot{x}_i]=V^2 s_i, \qquad \left[\frac{\partial F_{ij}}{\partial X_k}\right]=s_i F_{mj} F_{lk} n_m n_l; \tag{2.11}$$

here $\boldsymbol{s}$ is a vector which may be called the *amplitude*, V is the speed of propagation, and $\boldsymbol{n}$ is the *unit normal* to the singular surface. We call $\boldsymbol{n}$ the *direction of the wave.*

In terms of the Piola-Kirchhoff tensor, the dynamical equations take the form

$$\frac{\partial S_{ij}}{\partial X_j}+b_i=\ddot{x}_i \tag{2.12}$$

---

* We assume that the reader has some familiarity with the kinematics of singular surfaces. A thorough exposition is that of TRUESDELL & TOUPIN [*5*], §§ 180—191.

** See Remark 4.1 of II. Other transparent conditions on the motion sufficient for A-2 are given in § 5 of I.

*** *Vid.* TRUESDELL & TOUPIN [*5*], § 190, who attribute (2.11) to HUGONIOT and HADAMARD.

on either side of the wave. We shall soon show that our hypotheses A-1 and A-2 and our smoothness assumption about $\mathfrak{S}$ guarantee the existence and continuity in $\boldsymbol{X}$ of the term $\operatorname{div}_{\boldsymbol{X}}\boldsymbol{S}$ in (2.12) for $\boldsymbol{X}$ not on the singular surface. Of course, by A-1, $\ddot{\boldsymbol{x}}$ has only a jump discontinuity at that surface. Hence, (2.12) and (2.10) yield

$$\left[\frac{\partial S_{ij}}{\partial X_j}\right] = [\ddot{x}_i]. \tag{2.13}$$

When the time t is fixed, $\boldsymbol{F}$ and $\boldsymbol{S}$ are tensor-valued fields, while $\boldsymbol{F}_r^t$ and $\boldsymbol{F}^t$ are function-valued fields over the region $\mathscr{B}$ occupied by the body in its reference configuration. The values $\boldsymbol{F}_r^t(\boldsymbol{X}, \cdot)$, $\boldsymbol{F}^t(\boldsymbol{X}, \cdot)$ of the fields $\boldsymbol{F}_r^t$, $\boldsymbol{F}^t$ are in $\mathscr{H}$ and $\mathscr{H}^{\oplus}$ respectively. In A-1 we assume that at each point not on the wave the gradient $\nabla_{\boldsymbol{X}}\boldsymbol{F}$ of the field $\boldsymbol{F}$ exists and is continuous. In A-2 we assume that the gradient $\nabla_{\boldsymbol{X}}\boldsymbol{F}_r^t$ of $\boldsymbol{F}_r^t$ exists, is in $\mathscr{H}$, and is continuous at all points. Hence, the gradient $\nabla_{\boldsymbol{X}}\boldsymbol{F}^t$ of $\boldsymbol{F}^t$ exists, is in $\mathscr{H}^{\oplus}$, and is a continuous field over $\mathscr{B}$, except on the wave. It now follows from our smoothness hypothesis on $\mathfrak{S}$ that $\nabla_{\boldsymbol{X}}\boldsymbol{S}$ (and hence $\operatorname{div}_{\boldsymbol{X}}\boldsymbol{S}$ mentioned earlier) exist and are continuous in $\boldsymbol{X}$ before and after the wave. Furthermore a familiar chain rule and (2.6) yield the following expressions which are valid at all points not on the singular surface:

$$\nabla_{\boldsymbol{X}}\boldsymbol{S} = \delta\mathfrak{S}(\boldsymbol{F}^t|\nabla_{\boldsymbol{X}}\boldsymbol{F}^t), \tag{2.14}$$

$$\nabla_{\boldsymbol{X}}\boldsymbol{S} = D_{\boldsymbol{F}}\mathfrak{S}(\boldsymbol{F}^t|\nabla_{\boldsymbol{X}}\boldsymbol{F}) + \delta_{\boldsymbol{F}_r^t}\mathfrak{S}(\boldsymbol{F}^t|\nabla_{\boldsymbol{X}}\boldsymbol{F}_r^t). \tag{2.15}$$

Since $\nabla_{\boldsymbol{X}}\boldsymbol{S}(\boldsymbol{X}, t)$ is really a third-order tensor, *i.e.*, a linear transformation which operates on a vector $\boldsymbol{v}$ to give the second-order tensor $\nabla_{\boldsymbol{X}}\boldsymbol{S}(\boldsymbol{X}, t)\{\boldsymbol{v}\}$, and $\nabla_{\boldsymbol{X}}\boldsymbol{F}^t(\boldsymbol{X}, \cdot)$ is really a linear transformation which operates on a vector $\boldsymbol{v}$ to give the function $\nabla_{\boldsymbol{X}}\boldsymbol{F}^t(\boldsymbol{X}, \cdot)\{\boldsymbol{v}\}$ in $\mathscr{H}^{\oplus}$, it would be more precise to replace equation (2.14) by the assertion that

$$\nabla_{\boldsymbol{X}}\boldsymbol{S}(\boldsymbol{X}, t)\{\boldsymbol{v}\} = \delta\mathfrak{S}(\boldsymbol{F}^t(\boldsymbol{X}, t)|\nabla_{\boldsymbol{X}}\boldsymbol{F}^t(\boldsymbol{X}, \cdot)\{\boldsymbol{v}\})$$

for all vectors $\boldsymbol{v}$. A notation of such precision would, however, obscure rather than clarify the main ideas of our argument. (Similar remarks apply to equations (2.15), (2.16), (2.21), (2.22), and (3.10).)

It of course follows from A-1 and A-2 that $\boldsymbol{F}^t$ and $\nabla_{\boldsymbol{X}}\boldsymbol{F}_r^t$, as points in their appropriate spaces, are both continuous across the wave. Hence, since $\delta\mathfrak{S}(\cdot|\cdot)$ is jointly continuous in both its variables, $\delta_{\boldsymbol{F}_r^t}\mathfrak{S}(\boldsymbol{F}^t|\nabla_{\boldsymbol{X}}\boldsymbol{F}_r^t)$ is continuous across the wave, and (2.15) yields the following expression for the jump in $\nabla_{\boldsymbol{X}}\boldsymbol{S}$:

$$[\nabla_{\boldsymbol{X}}\boldsymbol{S}] = [D_{\boldsymbol{F}}\mathfrak{S}(\boldsymbol{F}^t|\nabla_{\boldsymbol{X}}\boldsymbol{F})]. \tag{2.16}$$

Using (2.7), the continuity of $\mathfrak{S}_{pq}(\boldsymbol{F}^t)$ in $\boldsymbol{F}^t$, and the continuity of $\boldsymbol{F}^t$ across the wave, we can write (2.16) as follows:

$$\left[\frac{\partial S_{ij}}{\partial X_k}\right] = D_{F_{mp}}\mathfrak{S}_{ij}(\boldsymbol{F}^t)\left[\frac{\partial F_{mp}}{\partial X_k}\right]. \tag{2.17}$$

Here $\boldsymbol{F}^t$ is the history of the deformation gradient at the wave.

Substituting (2.17) and (2.11) into (2.13), we obtain the following generalization* of the Fresnel-Hadamard theorem.

* As we mentioned in the preface, propagation conditions of the type (2.18) are known in several special theories; for elasticity theory (2.18) was derived by Hadamard [9] and for linear viscoelasticity by Herrera & Gurtin [11]. We have recently seen a manuscript by Varley [12] in which he arrives at (2.18) for materials of integral type. The important point of Theorem 2.1 is not that an equation of the form (2.18) holds, but rather that the accoustic tensor $\boldsymbol{Q}(\boldsymbol{n})$ in (2.18) is given in terms of instantaneous derivatives by the *general formula* (2.9).

Theorem 2.1. *The amplitude* $\mathbf{s}$ *and the speed of propagation* V *of an acceleration wave traveling in the direction* $\boldsymbol{n}$ *in a material of the type* (2.2) *must satisfy the propagation condition*

$$\underset{\sim}{Q}(\boldsymbol{n})\,\mathbf{s} = V^2\mathbf{s}, \tag{2.18}$$

where $\boldsymbol{Q}(\boldsymbol{n})$ *is the instantaneous acoustic tensor* (2.9) *corresponding to the history of the deformation gradient at the wave.*

Waves of Higher Order

Let $N \geqq 2$ be an integer. A singular surface of the motion $\boldsymbol{x}(\boldsymbol{X}, t)$ is called a *wave of order* N if the following two conditions hold:

B-1: $\boldsymbol{x}$ and its first $N-1$ derivatives are continuous in $\boldsymbol{X}$ and t jointly for all $\boldsymbol{X}$ and t, while all N^{th}-order derivatives of $\boldsymbol{x}$ suffer jump discontinuities across the surface, but are continuous in $\boldsymbol{X}$ and t jointly everywhere else.

B-2: For each fixed t the mapping $\boldsymbol{X} \to \boldsymbol{F}_r^t(\boldsymbol{X}, \cdot)$ with values in $\mathcal{H}$, is $N-1$ times continuously Fréchet-differentiable for all $\boldsymbol{X}$.

When concerned with a wave of order N, we assume that $\boldsymbol{b}(\boldsymbol{X}, t)$ is $N-1$ times differentiable with respect to $\boldsymbol{X}$ and that $\nabla_{\boldsymbol{X}}^{N-1}\boldsymbol{b}$ is continuous across the wave.

Whereas for our theory of acceleration waves it sufficed to assume that $\mathfrak{S}$ is once continuously differentiable over its domain $\mathcal{D}$, to discuss a wave of order $N > 2$ we must assume that $\mathfrak{S}$ is $N-1$ times continuously Fréchet-differentiable on $\mathcal{D}$.

Let us consider a wave of order 3. The following kinematical conditions of compatibility must hold for the jumps across a third-order singular surface with unit normal $\boldsymbol{n}$:*

$$\left[\frac{\partial \ddot{x}_i}{\partial X_k}\right] = V^2 s_i F_{mk} n_m, \qquad \left[\frac{\partial^2 F_{ij}}{\partial X_l\,\partial X_k}\right] = s_i F_{mj} F_{pk} F_{ql} n_m n_p n_q; \tag{2.19}$$

$\mathbf{s}$ is again the amplitude and V the speed of propagation. Of course (2.12) holds; B-1, B-2, and our smoothness assumptions on $\mathfrak{S}$ permit us to differentiate (2.12) at all points not on the wave:

$$\frac{\partial^2 S_{ij}}{\partial X_k\,\partial X_j} + \frac{\partial b_i}{\partial X_k} = \frac{\partial \ddot{x}_i}{\partial X_k}. \tag{2.20}$$

Taking the gradient with respect to $\boldsymbol{X}$ of (2.15), we find that

$$\begin{aligned} \nabla_{\boldsymbol{X}}^2 \boldsymbol{S} = {} & D_{\boldsymbol{F}}\mathfrak{S}(\boldsymbol{F}^t|\nabla_{\boldsymbol{X}}^2\boldsymbol{F}) + D_{\boldsymbol{F}}^2\mathfrak{S}(\boldsymbol{F}^t|\nabla_{\boldsymbol{X}}\boldsymbol{F}, \nabla_{\boldsymbol{X}}\boldsymbol{F}) + \delta_{\boldsymbol{F}_r^t} D_{\boldsymbol{F}}\mathfrak{S}(\boldsymbol{F}^t|\nabla_{\boldsymbol{X}}\boldsymbol{F}, \nabla_{\boldsymbol{X}}\boldsymbol{F}_r^t) + \\ & + D_{\boldsymbol{F}}\,\delta_{\boldsymbol{F}_r^t}\mathfrak{S}(\boldsymbol{F}^t|\nabla_{\boldsymbol{X}}\boldsymbol{F}_r^t, \nabla_{\boldsymbol{X}}\boldsymbol{F}) + \delta_{\boldsymbol{F}_r^t}^2\mathfrak{S}(\boldsymbol{F}^t|\nabla_{\boldsymbol{X}}\boldsymbol{F}_r^t, \nabla_{\boldsymbol{X}}\boldsymbol{F}_r^t) + \delta_{\boldsymbol{F}_r^t}\mathfrak{S}(\boldsymbol{F}^t|\nabla_{\boldsymbol{X}}^2\boldsymbol{F}_r^t); \end{aligned} \tag{2.21}$$

$D_{\boldsymbol{F}}^2\mathfrak{S}$ is defined by replacing $\mathfrak{S}$ in (2.5a) with $D_{\boldsymbol{F}}\mathfrak{S}$; $\delta_{\boldsymbol{F}_r^t} D_{\boldsymbol{F}}\mathfrak{S}$ is defined by replacing $\mathfrak{S}$ in (2.5b) with $D_{\boldsymbol{F}}\mathfrak{S}$; *etc.*** It follows from B-1, B-2, and the continuity of the first two Fréchet derivatives of $\mathfrak{S}$ that all of the terms on the right in (2.21), except the first, are continuous across the wave. The first term suffers

* *Vid.* TRUESDELL & TOUPIN [5], § 191; they attribute the general kinematical conditions of compatibility for waves of order $N > 2$ to HADAMARD.

** *Cf.* eq. (1.7) of II; we note that if $\boldsymbol{F}^t$ is in $\mathcal{D}$, $\boldsymbol{G}_r^t$ in $\mathcal{H}$, and $\boldsymbol{A}$ any tensor, then $\delta_{\boldsymbol{F}_r^t} D_{\boldsymbol{F}}\mathfrak{S}(\boldsymbol{F}^t|\boldsymbol{A}, \boldsymbol{G}_r^t) = D_{\boldsymbol{F}}\,\delta_{\boldsymbol{F}_r^t}\mathfrak{S}(\boldsymbol{F}^t|\boldsymbol{G}_r^t, \boldsymbol{A})$.

a jump discontinuity at the wave but is continuous everywhere else. Hence

$$[\nabla_{\boldsymbol{X}}^2 \boldsymbol{S}] = [D_{\boldsymbol{F}} \mathfrak{S}(\boldsymbol{F}^t | \nabla_{\boldsymbol{X}}^2 \boldsymbol{F})], \tag{2.22}$$

i.e.,

$$\left[\frac{\partial S_{ij}}{\partial X_k \, \partial X_l}\right] = D_{F_{mp}} \mathfrak{S}_{ij}(\boldsymbol{F}^t) \left[\frac{\partial^2 F_{mp}}{\partial X_k \, \partial X_l}\right], \tag{2.23}$$

and, since $\nabla_{\boldsymbol{X}} \boldsymbol{b}$ is assumed continuous, (2.20) yields

$$D_{F_{mp}} \mathfrak{S}_{ij}(\boldsymbol{F}^t) \left[\frac{\partial^2 F_{mp}}{\partial X_k \, \partial X_j}\right] = \left[\frac{\partial \ddot{x}_i}{\partial X_k}\right]. \tag{2.24}$$

From (2.19) and (2.24) we obtain

$$D_{F_{mp}} \mathfrak{S}_{ij}(\boldsymbol{F}^t) \, F_{ap} F_{bj} F_{ck} n_a n_b n_c s_m = V^2 s_i F_{mk} n_m, \tag{2.25}$$

which, by (2.9), is equivalent to the propagation condition (2.18). Since similar analyses may be applied to waves of order greater than 3, we have the following extension of the Ericksen-Truesdell theorem to general materials with memory.

Theorem 2.2. *Theorem 2.1 holds not only for acceleration waves but also for all waves of order $N \geqq 2$.*

Thus, the tensor $\boldsymbol{Q}(\boldsymbol{n})$ defined in (2.9) is the true acoustic tensor for all waves of order $N \geqq 2$.*

3. Homothermal and Homentropic Acceleration Waves

We now allow the stress tensor to depend not only on the history of the deformation gradient but also on the history of a thermodynamic variable such as the temperature ϑ or the entropy η.

Homothermal Waves

Let us first take the temperature as an independent variable and write

$$\boldsymbol{S}(t) = \bar{\mathfrak{S}}(\boldsymbol{F}^t, \vartheta^t), \tag{3.1}$$

where $\boldsymbol{F}^t$ is defined in (2.1), and ϑ^t, called the *temperature history*, is given by

$$\vartheta^t(s) = \vartheta(\boldsymbol{X}, t - s), \qquad 0 \leqq s < \infty. \tag{3.2}$$

The restriction ϑ_r^t of ϑ^t to $(0, \infty)$ is called the *past history of the temperature.* Our *smoothness hypothesis on the constitutive functional* $\bar{\mathfrak{S}}$ of (3.1) is an obvious extension of that imposed on $\mathfrak{S}$ in the previous section. We suppose that the histories $\boldsymbol{F}^t$, ϑ^t belong to function spaces $\mathfrak{H}^\oplus, \mathscr{H}^\oplus$ with norms of the type (1.2), and consider the product space $\mathscr{S} = \mathfrak{H}^\oplus \times \mathscr{H}^\oplus$ with the direct sum norm defined in (1.3). We assume that the domain of definition of $\bar{\mathfrak{S}}$ is an open subset $\bar{\mathscr{D}}$ of $\mathscr{S}$ and that $\bar{\mathfrak{S}}$ is continuously Fréchet-differentiable throughout $\bar{\mathscr{D}}$. For the results of this section there is no need to specify $\|\cdot\|$ in (1.2); *i.e.*, the norms over the spaces $\mathfrak{H}$, $\mathscr{H}$ of past histories $\boldsymbol{F}_r^t$, ϑ_r^t may be left arbitrary.

---

* Although all weak waves obey the same propagation condition, in view of recent researches of Green [*18*] on waves in isotropic elastic materials, we expect that the laws of growth and decay for waves with $N > 2$ should be simpler than for those with $N = 2$.

The argument which previously gave us (2.4) now permits us to write

$$S = \overline{\mathfrak{S}}(F_r^t, \vartheta_r^t; F, \vartheta) \tag{3.3}$$

with the abbreviations $S = S(t)$, $F = F(t)$, $\vartheta = \vartheta(t)$.

Our smoothness hypothesis on the functional $\overline{\mathfrak{S}}$ of (3.1) insures the existence of the following partial derivatives for every pair (F^t, ϑ^t) in $\overline{\mathscr{D}}$, every pair (G_r, g_r) in $\mathscr{H} \times \mathscr{H}$, every tensor A, and every scalar α:

$$D_F \overline{\mathfrak{S}}(F^t, \vartheta^t | A) \stackrel{\text{def}}{=} \frac{\partial}{\partial \lambda} \overline{\mathfrak{S}}(F_r^t, \vartheta_r^t; F + \lambda A, \vartheta)\Big|_{\lambda=0}, \tag{3.4a}$$

$$\delta_{F_r^t} \overline{\mathfrak{S}}(F^t, \vartheta^t | G_r) \stackrel{\text{def}}{=} \frac{\partial}{\partial \lambda} \overline{\mathfrak{S}}(F_r^t + \lambda G_r, \vartheta_r^t; F, \vartheta)\Big|_{\lambda=0}, \tag{3.4b}$$

$$D_\vartheta \overline{\mathfrak{S}}(F^t, \vartheta^t | \alpha) \stackrel{\text{def}}{=} \frac{\partial}{\partial \lambda} \overline{\mathfrak{S}}(F_r^t, \vartheta_r^t; F, \vartheta + \lambda \alpha)\Big|_{\lambda=0}, \tag{3.4c}$$

$$\delta_{\vartheta_r^t} \overline{\mathfrak{S}}(F^t, \vartheta | g_r) \stackrel{\text{def}}{=} \frac{\partial}{\partial \lambda} \overline{\mathfrak{S}}(F_r^t, \vartheta_r^t + \lambda g_r; F, \vartheta)\Big|_{\lambda=0}. \tag{3.4d}$$

Each of the functionals defined in (3.4) is jointly continuous in all of its arguments and linear in the argument after the stroke. These functionals, being, respectively, the Fréchet-derivatives of the mappings $F \to \overline{\mathfrak{S}}(F_r^t, \vartheta_r^t; F, \vartheta)$, $F_r^t \to \overline{\mathfrak{S}}(F_r^t, \vartheta_r^t; F, \vartheta)$, $\vartheta \to \overline{\mathfrak{S}}(F_r^t, \vartheta_r^t; F, \vartheta)$, and $\vartheta_r^t \to \overline{\mathfrak{S}}(F_r^t, \vartheta_r^t; F, \vartheta)$, are related to the Fréchet-derivative $\delta \overline{\mathfrak{S}}(F^t, \vartheta^t | G, g)$ of the mapping $(F^t, \vartheta^t) \to \overline{\mathfrak{S}}(F^t, \vartheta^t)$ by the equation

$$\begin{aligned} \delta \overline{\mathfrak{S}}(F^t, \vartheta^t | G, g) = D_F \overline{\mathfrak{S}}(F^t, \vartheta^t | G(0)) + \delta_{F_r^t} \overline{\mathfrak{S}}(F^t, \vartheta^t | G_r) + \\ + D_\vartheta \overline{\mathfrak{S}}(F^t, \vartheta^t | g(0)) + \delta_{\vartheta_r^t} \overline{\mathfrak{S}}(F^t, \vartheta^t | g_r), \end{aligned} \tag{3.5}$$

which holds for every function-pair (G, g) in $\mathscr{H}^\oplus \times \mathscr{H}^\oplus$.

Following (2.7), we let $D_{F_{ij}} \overline{\mathfrak{S}}_{pq}(F^t, \vartheta^t)$ be the Cartesian components of the linear transformation mapping the tensor A into the tensor $D_F \overline{\mathfrak{S}}(F^t, \vartheta^t | A)$:

$$D_{F_{ij}} \overline{\mathfrak{S}}_{pq}(F^t, \vartheta^t) A_{ij} = [D_F \overline{\mathfrak{S}}(F^t, \vartheta^t | A)]_{pq}. \tag{3.6}$$

For convenience, let us put

$$D_\vartheta \overline{\mathfrak{S}}(F^t, \vartheta^t) \stackrel{\text{def}}{=} D_\vartheta \overline{\mathfrak{S}}(F^t, \vartheta^t | 1); \tag{3.7a}$$

then, by the linearity of $D_\vartheta \overline{\mathfrak{S}}(F^t, \vartheta^t | \cdot)$, we have

$$D_\vartheta \overline{\mathfrak{S}}(F^t, \vartheta^t | \alpha) = \alpha D_\vartheta \overline{\mathfrak{S}}(F^t, \vartheta^t). \tag{3.7b}$$

Of course, $D_\vartheta \overline{\mathfrak{S}}(F^t, \vartheta^t)$ is a second-order tensor; for its components we write $D_\vartheta \overline{\mathfrak{S}}_{pq}(F^t, \vartheta)$. It follows from (3.4), (3.6), and (3.7) that

$$D_{F_{ij}} \overline{\mathfrak{S}}_{pq}(F^t, \vartheta^t) = \frac{\partial}{\partial F_{ij}} \overline{\mathfrak{S}}_{pq}(F_r^t, \vartheta_r^t; F, \vartheta) \tag{3.8a}$$

and

$$D_\vartheta \overline{\mathfrak{S}}_{pq}(F^t, \vartheta^t) = \frac{\partial}{\partial \vartheta} \overline{\mathfrak{S}}_{pq}(F_r^t, \vartheta_r^t; F, \vartheta). \tag{3.8b}$$

Hence $D_{F_{ij}} \overline{\mathfrak{S}}_{pq}(F^t, \vartheta^t)$ is simply the partial derivative of $\overline{\mathfrak{S}}_{pq}(F_r^t, \vartheta_r^t; F, \vartheta)$ with respect to the component F_{ij} of the present value F; this derivative is taken holding fixed the past history F_r^t and the (entire) history ϑ^t. An analogous interpretation applies to $D_\vartheta \overline{\mathfrak{S}}_{pq}(F^t, \vartheta^t)$.

Given any unit vector $\boldsymbol{n}$ and any pair $\boldsymbol{F}^t, \vartheta^t$ in $\overline{\mathscr{D}}$, we call the tensor $\overline{\boldsymbol{Q}}(\boldsymbol{n})$ with components

$$\overline{Q}_{ij}(\boldsymbol{n}) = D_{F_{jm}} \overline{\mathfrak{S}}_{il}(\boldsymbol{F}^t, \vartheta^t)\, F_{pm} F_{ql} n_p n_q \tag{3.9}$$

the *instantaneous homothermal acoustic tensor* corresponding to $\boldsymbol{F}^t$, ϑ^t, and $\boldsymbol{n}$. Of course we have yet to show that this tensor really serves as an acoustic tensor for homothermal waves.

It is clear that the presence of ϑ^t in (3.1) requires us to strengthen our previous definition of an acceleration wave by laying down an hypothesis of continuity for the present temperature and an hypothesis restricting the wildness of the past variations of temperature. Thus, we here call a singular surface an *acceleration wave* if, in addition to the conditions A-1 and A-2 of Section 2, the following two hypotheses are fulfilled:

A-1′: ϑ is a continuous function of $\boldsymbol{X}$ and t jointly for all $\boldsymbol{X}$ and t, while $\dot{\vartheta}$ and $\nabla_{\boldsymbol{X}} \vartheta$ $\left(\text{with components } \frac{\partial \vartheta}{\partial X_j}\right)$ suffer at most jump discontinuities across the surface but are continuous functions of $\boldsymbol{X}$ and t jointly everywhere else.

A-2′: At each time t the mapping $\boldsymbol{X} \to \vartheta_r^t(\boldsymbol{X}, \cdot)$, with values in the function space $\mathscr{H}$, is (once) continuously Fréchet-differentiable.

Of course we also assume (2.10) and that for each $\boldsymbol{X}$ and t the pair $\left(\boldsymbol{F}^t(\boldsymbol{X}, \cdot), \vartheta^t(\boldsymbol{X}, \cdot)\right)$ belongs to the domain $\overline{\mathscr{D}}$ of $\overline{\mathfrak{S}}$.

The equations (2.11)—(2.13) hold again under our present assumptions. Use of A-1, A-2, A-1′, A-2′, and (3.5) now gives us, in place of (2.15), the following more complicated expression for $\nabla_{\boldsymbol{X}} \boldsymbol{S}$ at points near, but not on, the singular surface:

$$\begin{aligned} \nabla_{\boldsymbol{X}} \boldsymbol{S} = {} & D_{\boldsymbol{F}} \overline{\mathfrak{S}}(\boldsymbol{F}^t, \vartheta^t | \nabla_{\boldsymbol{X}} \boldsymbol{F}) + \delta_{\boldsymbol{F}_r^t} \overline{\mathfrak{S}}(\boldsymbol{F}^t, \vartheta^t | \nabla_{\boldsymbol{X}} \boldsymbol{F}_r^t) + \\ & + D_{\vartheta} \overline{\mathfrak{S}}(\boldsymbol{F}^t, \vartheta^t | \nabla_{\boldsymbol{X}} \vartheta) + \delta_{\vartheta_r^t} \overline{\mathfrak{S}}(\boldsymbol{F}^t, \vartheta^t | \nabla_{\boldsymbol{X}} \vartheta_r^t), \end{aligned} \tag{3.10}$$

where $\nabla_{\boldsymbol{X}} \boldsymbol{F}_r^t$ and $\nabla_{\boldsymbol{X}} \vartheta_r^t$ are the gradients of the function-valued fields $\boldsymbol{F}_r^t(\boldsymbol{X}, \cdot)$ and $\vartheta_r^t(\boldsymbol{X}, \cdot)$. The argument which previously gave us (2.17) now tells us that our smoothness hypotheses on $\overline{\mathfrak{S}}$ and the wave imply

$$\left[\frac{\partial S_{ij}}{\partial X_k}\right] = D_{F_{mp}} \overline{\mathfrak{S}}_{ij}(\boldsymbol{F}^t, \vartheta^t) \left[\frac{\partial F_{mp}}{\partial X_k}\right] + D_{\vartheta} \overline{\mathfrak{S}}_{ij}(\boldsymbol{F}^t, \vartheta^t) \left[\frac{\partial \vartheta}{\partial X_k}\right], \tag{3.11}$$

where $\boldsymbol{F}^t$ and ϑ^t are the histories of the deformation gradient and temperature of a particle situated on the wave at time t.

We call an acceleration wave *homothermal* if

$$[\nabla_{\boldsymbol{X}} \vartheta] = \boldsymbol{0}, \qquad [\dot{\vartheta}] = 0; \tag{3.12}$$

i.e., if $\nabla_{\boldsymbol{X}} \vartheta$ and $\dot{\vartheta}$ are continuous across the wave.

For a homothermal acceleration wave, (3.11) reduces to an equation of the form (2.17):

$$\left[\frac{\partial S_{ij}}{\partial X_k}\right] = D_{F_{mp}} \overline{\mathfrak{S}}_{ij}(\boldsymbol{F}^t, \vartheta^t) \left[\frac{\partial F_{mp}}{\partial X_k}\right]. \tag{3.13}$$

Substituting this result and (2.11) into (2.13), we obtain the following generalization of part of the first theorem of Duhem.

Theorem 3.1. *The amplitude* $\mathbf{s}$ *and the speed of propagation* V *of a homothermal acceleration wave traveling in the direction* $\boldsymbol{n}$ *in a general material of the type* (3.1) *must satisfy the propagation condition*

$$\bar{\boldsymbol{Q}}(\boldsymbol{n})\,\mathbf{s} = V^2\mathbf{s}, \tag{3.14}$$

where $\bar{\boldsymbol{Q}}(\boldsymbol{n})$ *is the instantaneous homothermal acoustic tensor* (3.9) *corresponding to the histories* $\boldsymbol{F}^t, \vartheta^t$ *at the wave.*

Thus, the tensor $\bar{\boldsymbol{Q}}(\boldsymbol{n})$ defined in (3.9) *is* an acoustic tensor for homothermal waves, and the name we gave it is justified.

Homentropic Acceleration Waves

Let us now take entropy as an independent thermodynamic variable and write

$$\boldsymbol{S}(t) = \hat{\mathfrak{S}}(\boldsymbol{F}^t, \eta^t) \tag{3.15}$$

with $\boldsymbol{F}^t$ defined in (2.1) and η^t, called the *entropy history*, given by

$$\eta^t(s) = \eta(\boldsymbol{X}, t-s), \qquad 0 \leqq s < \infty. \tag{3.16}$$

The theory of propagation conditions for homentropic waves parallels perfectly the theory just given for homothermal waves. In fact, we obtain the homentropic theory from the homothermal theory by making the following replacements in the preceding subsection: ϑ by η, $\bar{\mathfrak{S}}$ by $\hat{\mathfrak{S}}$, the word *temperature* by *entropy*, and the word *homothermal* by *homentropic*. Thus, we call the tensor $\hat{\boldsymbol{Q}}(\boldsymbol{n})$, with components

$$\hat{Q}_{ij}(\boldsymbol{n}) = D_{F_{jm}}\hat{\mathfrak{S}}_{il}(\boldsymbol{F}^t, \eta^t)\, F_{pm} F_{ql} n_p n_q, \tag{3.17}$$

the *instantaneous homentropic acoustic tensor* corresponding to $\boldsymbol{F}^t, \eta^t$, and $\boldsymbol{n}$, and we read off the following counterpart of Theorem 3.1.

Theorem 3.2. *The amplitude* $\mathbf{s}$ *and the speed of propagation* V *of a homentropic acceleration wave traveling in the direction* $\boldsymbol{n}$ *in a general material of the type* (3.15) *must satisfy the propagation condition*

$$\hat{\boldsymbol{Q}}(\boldsymbol{n})\,\mathbf{s} = V^2\mathbf{s}, \tag{3.18}$$

where $\hat{\boldsymbol{Q}}(\boldsymbol{n})$ *is the instantaneous homentropic acoustic tensor* (3.17) *corresponding to the histories* $\boldsymbol{F}^t$, η^t *at the wave.*

This completes our program of extending the first theorem of DUHEM to general materials with memory.

Of main interest here is the observation that, in spite of the extreme generality of our starting hypotheses, the tensors $\bar{\boldsymbol{Q}}(\boldsymbol{n})$ of (3.14) and $\hat{\boldsymbol{Q}}(\boldsymbol{n})$ of (3.18) are related to the stress functionals $\bar{\mathfrak{S}}$ and $\hat{\mathfrak{S}}$ through the very simple and transparent formulae (3.9) and (3.17).

4. Results from the Thermodynamics of Materials with Memory

In the remaining sections we shall explore the consequences for acceleration waves of the theory of the thermodynamics of materials with memory [7, 8].

In accordance with that theory we assume that the specific Helmholz free energy $\psi(t)$, the Piola-Kirchhoff stress tensor $\boldsymbol{S}(t)$, and the specific entropy $\eta(t)$ are determined at a point whenever the histories $\boldsymbol{F}^t$ and ϑ^t are specified; the heat flux $\boldsymbol{q}(t)$ is determined when, in addition to the histories $\boldsymbol{F}^t$, ϑ^t, the present value of the temperature gradient

$$\boldsymbol{g}(t) = \nabla_{\boldsymbol{x}} \vartheta(\boldsymbol{x}, t) \tag{4.1}$$

is given. Of course, $\boldsymbol{g}(t)$, just like $\boldsymbol{F}^t$ and ϑ^t, depends on the particle as well as the time. Since we usually regard ϑ as a function of $\boldsymbol{X}$ and t, the following formula for $\boldsymbol{g}$ is worth noting:

$$\boldsymbol{g}(\boldsymbol{X}, t) = \boldsymbol{F}^{T^{-1}}(\boldsymbol{X}, t) \nabla_{\boldsymbol{X}} \vartheta(\boldsymbol{X}, t). \tag{4.2}$$

Our constitutive assumptions may be written

$$\psi(t) = \bar{\mathfrak{p}}(\boldsymbol{F}^t, \vartheta^t), \tag{4.3 a}$$

$$\boldsymbol{S}(t) = \bar{\mathfrak{S}}(\boldsymbol{F}^t, \vartheta^t), \tag{4.3 b}$$

$$\eta(t) = \bar{\mathfrak{h}}(\boldsymbol{F}^t, \vartheta^t), \tag{4.3 c}$$

$$\boldsymbol{q}(t) = \bar{\mathfrak{q}}(\boldsymbol{F}^t, \vartheta^t, \boldsymbol{g}(t)); \tag{4.3 d}$$

the functionals $\bar{\mathfrak{p}}$, $\bar{\mathfrak{S}}$, $\bar{\mathfrak{h}}$, and $\bar{\mathfrak{q}}$ characterize the *material* under consideration.

We continue to assume that the material is homogeneous and that a homogeneous configuration has been taken as reference. We also assume that the constitutive functionals $\bar{\mathfrak{p}}$, $\bar{\mathfrak{S}}$, and $\bar{\mathfrak{h}}$ in (4.3) obey the smoothness hypothesis already laid down for $\bar{\mathfrak{S}}$ in the previous section; *i.e.*, we suppose that $\bar{\mathfrak{p}}$ and $\bar{\mathfrak{h}}$, as well as $\bar{\mathfrak{S}}$, are defined and continuously Fréchet-differentiable on an open subset $\bar{\mathscr{D}}$ of $\mathscr{H}^{\oplus} \times \mathscr{H}^{\oplus}$. The constitutive functional $\bar{\mathfrak{q}}$ is assumed to be defined and continuously Fréchet-differentiable on $\bar{\mathscr{D}} \times \mathscr{V}$, where $\mathscr{V}$ is our underlying three-dimensional vector space. These *smoothness assumptions* enable us to apply the operators $D_{\boldsymbol{F}}$, D_ϑ, $\delta_{\boldsymbol{F}_r^t}$, $\delta_{\vartheta_r^t}$ to all of the constitutive functionals in (4.3); for example, we define $D_{\boldsymbol{F}}\bar{\mathfrak{p}}$, $D_\vartheta\bar{\mathfrak{p}}$, $D_{F_{ij}}\bar{\mathfrak{p}}$, $\delta_{\boldsymbol{F}_r^t}\bar{\mathfrak{p}}$, $\delta_{\vartheta_r^t}\bar{\mathfrak{p}}$ by replacement of $\mathfrak{S}$ with $\mathfrak{p}$ in equations (3.4)—(3.7).

Let r denote the specific extrinsic heat supply (with units of energy $\cdot$ time$^{-1}\cdot$ mass^{-1}) and ε the specific internal energy (with units of energy $\cdot$ mass^{-1}); ε is related to ψ by the familiar relation

$$\varepsilon = \psi + \vartheta\eta. \tag{4.4}$$

The law of *balance of energy* asserts that at each time and in each part $\mathscr{P}$ of the body

$$\frac{d}{dt}\int_{\mathscr{P}} \left(\frac{1}{2}\dot{\boldsymbol{x}}\cdot\dot{\boldsymbol{x}} + \varepsilon\right) dm = \int_{\mathscr{P}} (\dot{\boldsymbol{x}}\cdot\boldsymbol{b} + r)\, dm + \int_{\partial\mathscr{P}} (\dot{\boldsymbol{x}}\cdot\boldsymbol{T}\boldsymbol{n} - \boldsymbol{q}\cdot\boldsymbol{n})\, da, \tag{4.5}$$

where dm is the element of mass in the body, $\partial\mathscr{P}$ the surface of $\mathscr{P}$, and da and $\boldsymbol{n}$ the element of surface area and unit normal vector for $\partial\mathscr{P}$ in the configuration at time t.

In the terminology of Coleman & Noll [*13*], a *thermodynamic process* is a specification of the eight fields $\boldsymbol{x}$, $\boldsymbol{T}$ (with $\boldsymbol{T} = \boldsymbol{T}^T$), $\boldsymbol{b}$, ε, η, ϑ (with $\vartheta > 0$), $\boldsymbol{q}$, and r, as functions of $\boldsymbol{X}$ and t, compatible with the laws of balance of momentum and balance of energy. A thermodynamic process is said to be *admissible* in the body under consideration if it is compatible with the constitutive assumptions made about the body; *i.e.*, if it is compatible with (4.3).

The Second Law of Thermodynamics asserts that in every thermodynamic process the following *Clausius-Duhem inequality* must hold at each time and in each part $\mathscr{P}$ of the body:

$$\frac{d}{dt}\int_{\mathscr{P}} \eta\, dm \geqq \int_{\mathscr{P}} \frac{r}{\vartheta}\, dm - \int_{\partial\mathscr{P}} \frac{1}{\vartheta}\, \boldsymbol{q}\cdot\boldsymbol{n}\, da. \tag{4.6}$$

Theorem 4.1.* *If the Clausius-Duhem inequality* (4.6) *holds for every smooth admissible process, and if the norms on the function spaces* $\mathfrak{H}$ *and* $\mathscr{H}$ *are equal to the h-norm* (1.4) *of the theory of fading memory, then* $\bar{\mathfrak{p}}$ *must determine* $\bar{\mathfrak{S}}$ *through the* ***stress relation***

$$\bar{\mathfrak{S}}_{ij}(\boldsymbol{F}^t, \vartheta^t) = D_{F_{ij}} \bar{\mathfrak{p}}(\boldsymbol{F}^t, \vartheta^t) \tag{4.7}$$

and $\bar{\mathfrak{h}}$ *through the* ***entropy relation***

$$\bar{\mathfrak{h}}(\boldsymbol{F}^t, \vartheta^t) = -D_\vartheta \bar{\mathfrak{p}}(\boldsymbol{F}^t, \vartheta^t). \tag{4.8}$$

Since we have assumed that $\bar{\mathfrak{p}}$, $\bar{\mathfrak{S}}$, and $\bar{\mathfrak{h}}$ are all continuously Fréchet-differentiable throughout their common domain $\bar{\mathscr{D}}$, it follows from (4.7) and (4.8) that we can define second-order derivatives such as $D_{\boldsymbol{F}}^2\bar{\mathfrak{p}}$, $D_{\boldsymbol{F}} D_\vartheta \bar{\mathfrak{p}}$, $D_\vartheta D_{\boldsymbol{F}} \bar{\mathfrak{p}}$ and be sure that these derivatives exist, are jointly continuous in all their variables, and obey formulae of the type:

$$\begin{aligned}
D_{\boldsymbol{F}}^2 \mathfrak{p}(\boldsymbol{F}^t, \vartheta^t | \boldsymbol{A}, \boldsymbol{B}) &= D_{F_{ij}} D_{F_{kl}} \mathfrak{p}(\boldsymbol{F}^t, \vartheta^t)\, A_{ij} B_{kl} \\
&= \frac{\partial^2}{\partial\nu\,\partial\lambda}\, \mathfrak{p}(\boldsymbol{F}_r^t, \vartheta_r^t; \boldsymbol{F} + \nu\boldsymbol{A} + \lambda\boldsymbol{B}, \vartheta)\Big|_{\substack{\lambda=0\\ \nu=0}}, \\
D_{\boldsymbol{F}} D_\vartheta \mathfrak{p}(\boldsymbol{F}^t, \vartheta^t | \boldsymbol{A}, \alpha) &= D_{F_{ij}} D_\vartheta \mathfrak{p}(\boldsymbol{F}^t, \vartheta^t)\, \alpha A_{ij} \\
&= \frac{\partial^2}{\partial\nu\,\partial\lambda}\, \mathfrak{p}(\boldsymbol{F}_r^t, \vartheta_r^t; \boldsymbol{F} + \nu\boldsymbol{A}, \vartheta + \lambda\alpha)\Big|_{\substack{\lambda=0\\ \nu=0}}, \\
D_\vartheta D_{\boldsymbol{F}} \mathfrak{p}(\boldsymbol{F}^t, \vartheta^t | \boldsymbol{A}, \alpha) &= D_\vartheta D_{F_{ij}} \mathfrak{p}(\boldsymbol{F}^t, \vartheta^t)\, A_{ij}\alpha \\
&= \frac{\partial^2}{\partial\nu\,\partial\lambda}\, \mathfrak{p}(\boldsymbol{F}_r^t, \vartheta_r^t; \boldsymbol{F} + \nu\boldsymbol{A}, \vartheta + \lambda\alpha)\Big|_{\substack{\lambda=0\\ \nu=0}}.
\end{aligned} \tag{4.9}$$

The continuity of the functions on the right in (4.9) permits interchange of the order of differentiation with respect to λ and ν. In other words, *since* $\bar{\mathfrak{p}}$ *is twice continuously differentiable over* $\bar{\mathscr{D}}$, *the operators* $D_{F_{ij}}$, D_ϑ, *when applied to* $\bar{\mathfrak{p}}$ *two-at-a-time, commute with each other.* In particular, we have

$$D_{F_{ij}} D_{F_{kl}} \bar{\mathfrak{p}}(\boldsymbol{F}^t, \vartheta^t) = D_{F_{kl}} D_{F_{ij}} \bar{\mathfrak{p}}(\boldsymbol{F}^t, \vartheta^t), \tag{4.10}$$

and

$$D_{F_{ij}} D_\vartheta \bar{\mathfrak{p}}(\boldsymbol{F}^t, \vartheta^t) = D_\vartheta D_{F_{ij}} \bar{\mathfrak{p}}(\boldsymbol{F}^t, \vartheta^t). \tag{4.11}$$

It follows from (4.7) and (4.10) that

$$D_{F_{ij}} \bar{\mathfrak{S}}_{pq}(\boldsymbol{F}^t, \vartheta^t) = D_{F_{ij}} D_{F_{pq}} \bar{\mathfrak{p}}(\boldsymbol{F}^t, \vartheta^t) = D_{F_{pq}} D_{F_{ij}} \bar{\mathfrak{p}}(\boldsymbol{F}^t, \vartheta^t) = D_{F_{pq}} \bar{\mathfrak{S}}_{ij}(\boldsymbol{F}^t, \vartheta^t), \tag{4.12}$$

* COLEMAN [7], Theorems 1 and 2. When taken together, eqs. (4.5) and (4.6) are called "the generalized stress relation".

and, similarly, from (4.8) and (4.11) we conclude that

$$D_\vartheta \bar{\mathfrak{S}}_{ij}(\boldsymbol{F}^t, \vartheta^t) = - D_{F_{ij}} \bar{\mathfrak{h}}(\boldsymbol{F}^t, \vartheta^t). \tag{4.13}$$

When it is convenient to regard η^t rather than ϑ^t as an independent variable, in place of (4.3 a)—(4.3 d) we use the constitutive equations

$$\varepsilon(t) = \hat{\mathfrak{e}}(\boldsymbol{F}^t, \eta^t), \tag{4.14a}$$
$$\boldsymbol{S}(t) = \hat{\boldsymbol{\mathfrak{S}}}(\boldsymbol{F}^t, \eta^t), \tag{4.14b}$$
$$\vartheta(t) = \hat{\mathfrak{t}}(\boldsymbol{F}^t, \eta^t), \tag{4.14c}$$

and assume that the functionals $\hat{\mathfrak{e}}$, $\hat{\boldsymbol{\mathfrak{S}}}$, and $\hat{\mathfrak{t}}$ occurring here are defined and continuously Fréchet-differentiable on an open subset $\hat{\mathscr{D}}$ of $\mathscr{H}^\oplus \times \mathscr{H}^\oplus$.

In view of (4.4), the existence of the equations (4.14a)—(4.14c) is implied by (4.3 a)—(4.3 c) if, and only if, the functional transformation

$$(\boldsymbol{F}^t, \vartheta^t) \to (\boldsymbol{F}^t, \eta^t), \tag{4.15a}$$

defined by the functional $\bar{\mathfrak{h}}$, with

$$\eta^t(s) = \bar{\mathfrak{h}}(\boldsymbol{F}^{t-s}, \vartheta^{t-s}), \tag{4.15b}$$

is a *one-to-one* mapping of $\bar{\mathscr{D}}$ onto $\hat{\mathscr{D}}$. When this is the case, (4.14c) is the "inverse" of (4.3c),* and the following relation holds:

$$D_\eta \hat{\mathfrak{t}}(\boldsymbol{F}^t, \eta^t) = [D_\vartheta \bar{\mathfrak{h}}(\boldsymbol{F}^t, \vartheta^t)]^{-1}. \tag{4.16}$$

Assuming the invertibility of the functional transformation (4.15), one can prove**

Theorem 4.2. *The stress relation* (4.7) *and the entropy relation* (4.8) *imply that* $\hat{\mathfrak{e}}$ *determines* $\hat{\boldsymbol{\mathfrak{S}}}$ *through the* ***stress relation***

$$\hat{\mathfrak{S}}_{ij}(\boldsymbol{F}^t, \eta^t) = D_{F_{ij}} \hat{\mathfrak{e}}(\boldsymbol{F}^t, \eta^t) \tag{4.17}$$

and $\hat{\mathfrak{t}}$ *through the* ***temperature relation***

$$\hat{\mathfrak{t}}(\boldsymbol{F}^t, \eta^t) = D_\eta \hat{\mathfrak{e}}(\boldsymbol{F}^t, \eta^t). \tag{4.18}$$

The equations (4.17) and (4.18) imply the following counterparts of (4.10) and (4.11):

$$D_{F_{ij}} \hat{\mathfrak{S}}_{pq}(\boldsymbol{F}^t, \eta^t) = D_{F_{pq}} \hat{\mathfrak{S}}_{ij}(\boldsymbol{F}^t, \eta^t), \tag{4.19}$$
$$D_\eta \hat{\mathfrak{S}}_{ij}(\boldsymbol{F}^t, \eta^t) = D_{F_{ij}} \hat{\mathfrak{t}}(\boldsymbol{F}^t, \eta^t). \tag{4.20}$$

It follows from (4.3 a), (4.3 c), and (4.4) that $\varepsilon(t)$ is given by a functional $\bar{\mathfrak{e}}$ of $\boldsymbol{F}^t$ and ϑ^t:

$$\varepsilon(t) = \bar{\mathfrak{e}}(\boldsymbol{F}^t, \vartheta^t) = \bar{\mathfrak{p}}(\boldsymbol{F}^t, \vartheta^t) + \vartheta(t)\, \bar{\mathfrak{h}}(\boldsymbol{F}^t, \vartheta^t). \tag{4.21}$$

We call the number $\varkappa$ defined by

$$\varkappa = \varkappa(t) = D_\vartheta \bar{\mathfrak{e}}(\boldsymbol{F}^t, \vartheta^t) \tag{4.22}$$

* [8], § 11.
** COLEMAN [8], Theorem 6.

the *instantaneous heat capacity at fixed present strain* corresponding to $\boldsymbol{F}^t$ and ϑ; it measures the rate of change of internal energy with temperature at fixed strain in thermal impulses superimposed on $\boldsymbol{F}^t, \vartheta^t$ at time t. From (4.21), (4.22), (4.8), and (4.16) we conclude that

$$\varkappa(t) = \vartheta(t)\, D_\vartheta \overline{\mathfrak{h}}(\boldsymbol{F}^t, \vartheta^t) = \vartheta(t)\, [D_\eta \hat{\mathfrak{t}}(\boldsymbol{F}^t, \eta^t)]^{-1}. \tag{4.23}$$

If the material under consideration has been at rest at all past times with its present values of $\boldsymbol{F}$, ϑ, and η; then the histories $\boldsymbol{F}^t$, ϑ^t, η^t are the constant functions $\boldsymbol{F}^\dagger$, $\vartheta^\dagger$, $\eta^\dagger$ defined by

$$\boldsymbol{F}^\dagger(s) = \boldsymbol{F}(t) = \boldsymbol{F}, \qquad \vartheta^\dagger(s) = \vartheta(t) = \vartheta, \qquad \eta^\dagger(s) = \eta(t) = \eta, \qquad 0 \leqq s < \infty. \tag{4.24}$$

Let us consider the *equilibrium stress functions* $\overline{\boldsymbol{S}}^{(\infty)}$, $\hat{\boldsymbol{S}}^{(\infty)}$ giving the stress $\boldsymbol{S}^{(\infty)}$ corresponding to these constant histories:

$$\boldsymbol{S}^{(\infty)} = \overline{\boldsymbol{S}}^{(\infty)}(\boldsymbol{F}, \vartheta) \overset{\text{def}}{=} \overline{\mathfrak{S}}(\boldsymbol{F}^\dagger, \vartheta^\dagger), \tag{4.25a}$$

$$\boldsymbol{S}^{(\infty)} = \hat{\boldsymbol{S}}^{(\infty)}(\boldsymbol{F}, \eta) \overset{\text{def}}{=} \hat{\mathfrak{S}}(\boldsymbol{F}^\dagger, \eta^\dagger). \tag{4.25b}$$

Similarly, we may consider the *equilibrium energy function* $\bar{\varepsilon}^{(\infty)}$ with values $\varepsilon^{(\infty)}$

$$\varepsilon^{(\infty)} = \bar{\varepsilon}^{(\infty)}(\boldsymbol{F}, \vartheta) = \bar{\mathfrak{e}}(\boldsymbol{F}^\dagger, \vartheta^\dagger). \tag{4.26}$$

It should *not* be thought that *equilibrium* isothermal and adiabatic moduli of elasticity, defined by

$$\overline{B}^{(\infty)}_{ijpq} = \frac{\partial}{\partial F_{ik}} \overline{S}^{(\infty)}_{pl}(\boldsymbol{F}, \vartheta)\, F_{jk} F_{ql}, \tag{4.27a}$$

$$\hat{B}^{(\infty)}_{ijpq} = \frac{\partial}{\partial F_{ik}} \hat{S}^{(\infty)}_{pl}(\boldsymbol{F}, \eta)\, F_{jk} F_{ql}, \tag{4.27b}$$

and the *equilibrium* heat capacity, defined by

$$\varkappa^{(\infty)} = \frac{\partial}{\partial \vartheta} \bar{\varepsilon}^{(\infty)}(\boldsymbol{F}, \vartheta), \tag{4.28}$$

are, in general, equal respectively to the *instantaneous* isothermal and adiabatic moduli of elasticity,

$$\overline{B}^{(0)}_{ijpq} = D_{F_{ik}} \overline{\mathfrak{S}}_{pl}(\boldsymbol{F}^\dagger, \vartheta^\dagger)\, F_{jk} F_{ql}, \tag{4.29a}$$

$$\hat{B}^{(0)}_{ijpq} = D_{F_{ik}} \hat{\mathfrak{S}}_{pl}(\boldsymbol{F}^\dagger, \eta^\dagger)\, F_{jk} F_{ql}, \tag{4.29b}$$

and *heat capacity*

$$\varkappa^{(0)} = D_\vartheta \bar{\mathfrak{e}}(\boldsymbol{F}^\dagger, \vartheta^\dagger) \tag{4.30}$$

evaluated at the same rest histories. In the following remark, however, we see that the thermodynamics of materials with memory implies that, in a certain sense, the instantaneous heat capacity and elastic moduli evaluated at rest histories are "more positive" than the corresponding equilibrium quantities.

Remark 4.1. Let the hypothesis of Theorem 4.1 be assumed. Then for each pair $\boldsymbol{F}^\dagger$, $\vartheta^\dagger$ of constant histories in $\overline{\mathfrak{D}}$ we have

$$\varkappa^{(0)} \geqq \varkappa^{(\infty)} \tag{4.31}$$

and

$$\boldsymbol{C} = \boldsymbol{C}^T \Rightarrow \overline{B}^{(0)}_{ijpq} C_{ij} C_{pq} \geqq \overline{B}^{(\infty)}_{ijpq} C_{ij} C_{pq}. \tag{4.32}$$

Similarly, for each pair of constant histories $\boldsymbol{F}^\dagger$, $\eta^\dagger$ in $\hat{\mathfrak{D}}$

$$\boldsymbol{C} = \boldsymbol{C}^T \Rightarrow \hat{B}^{(0)}_{ijpq} C_{ij} C_{pq} \geqq \hat{B}^{(\infty)}_{ijpq} C_{ij} C_{pq}. \tag{4.33}$$

The conclusion (4.32) was obtained by COLEMAN in § 8 of [8].* To derive (4.33), one need merely replace $\bar{\mathfrak{p}}$ by $\bar{\mathfrak{e}}$ and ϑ^t by η^t throughout [8]; (4.31), on the other hand, is obtained by interchanging the roles of $\boldsymbol{F}^t$ and ϑ^t in § 8 of [8], with, of course, proper attention to difference in tensorial rank.

5. Properties of the Acoustic Tensor

Looking at (3.9), we see that

$$\begin{aligned}\bar{Q}_{ij}(\boldsymbol{n}) - \bar{Q}_{ji}(\boldsymbol{n}) &= [D_{F_{jm}} \bar{\mathfrak{S}}_{il}(\boldsymbol{F}^t, \vartheta^t) - D_{F_{im}} \bar{\mathfrak{S}}_{jl}(\boldsymbol{F}^t, \vartheta^t)]\, F_{pm} F_{ql} n_p n_q \\ &= [D_{F_{jm}} \bar{\mathfrak{S}}_{il}(\boldsymbol{F}^t, \vartheta^t) - D_{F_{il}} \bar{\mathfrak{S}}_{jm}(\boldsymbol{F}^t, \vartheta^t)]\, F_{pm} F_{ql} n_p n_q,\end{aligned} \tag{5.1}$$

and from (3.17) we get

$$\hat{Q}_{ij}(\boldsymbol{n}) - \hat{Q}_{ji}(\boldsymbol{n}) = [D_{F_{jm}} \hat{\mathfrak{S}}_{il}(\boldsymbol{F}^t, \eta^t) - D_{F_{il}} \hat{\mathfrak{S}}_{jm}(\boldsymbol{F}^t, \eta^t)]\, F_{pm} F_{ql} n_p n_q. \tag{5.2}$$

The equations (4.12) and (5.1) give $\bar{Q}_{ij}(\boldsymbol{n}) = \bar{Q}_{ji}(\boldsymbol{n})$, while (4.19) and (5.2) give $\hat{Q}_{ij}(\boldsymbol{n}) = \hat{Q}_{ji}(\boldsymbol{n})$. This proves our *main result***:

Theorem 5.1. *The stress relations* (4.12) *and* (4.19) *imply that the instantaneous homothermal acoustic tensor* (3.9) *and the instantaneous homentropic acoustic tensor* (3.17) *are both symmetric*:

$$\bar{\boldsymbol{Q}}(\boldsymbol{n}) = \bar{\boldsymbol{Q}}^T(\boldsymbol{n}), \tag{5.3}$$

$$\hat{\boldsymbol{Q}}(\boldsymbol{n}) = \hat{\boldsymbol{Q}}^T(\boldsymbol{n}). \tag{5.4}$$

The proper vectors of an acoustic tensor are called *acoustic axes*. It follows from (5.3) and (5.4) that for each homothermal or homentropic acceleration wave there exist at least three mutually orthogonal acoustic axes. We have not taken up the question of whether the corresponding speed V is real; in general, it need not be.***

We now seek a relation between the acoustic tensors (3.9) and (3.17). If we assume that the functional transformation (4.15) is smoothly invertible, a chain rule permits us to write

$$D_{F_{jm}} \hat{\mathfrak{S}}_{il}(\boldsymbol{F}^t, \eta^t) = D_{F_{jm}} \bar{\mathfrak{S}}_{il}(\boldsymbol{F}^t, \vartheta^t) + [D_\vartheta \bar{\mathfrak{S}}_{il}(\boldsymbol{F}^t, \vartheta^t)]\, [D_{F_{jm}} \mathfrak{t}(\boldsymbol{F}^t, \eta^t)]. \tag{5.5}$$

Since (4.20) and (4.23) with $\varkappa \neq 0$ give us

$$D_{F_{jm}} \mathfrak{t}(\boldsymbol{F}^t, \eta^t) = D_\eta \hat{\mathfrak{S}}_{jm}(\boldsymbol{F}^t, \eta^t) = D_\vartheta \bar{\mathfrak{S}}_{jm}(\boldsymbol{F}^t, \vartheta^t)\, D_\eta \mathfrak{t}(\boldsymbol{F}^t, \eta^t) = \frac{\vartheta}{\varkappa} D_\vartheta \bar{\mathfrak{S}}_{jm}(\boldsymbol{F}^t, \vartheta^t), \tag{5.6}$$

---

* Our present (4.32) is equivalent to (8.2) of [8].

It may be helpful to note that the eq. (8.3) of [8] is equivalent to the symmetry relation $\bar{B}^{(0)}_{ijpq} - \bar{B}^{(\infty)}_{ijpq} = \bar{B}^{(0)}_{pqij} - \bar{B}^{(\infty)}_{pqij}$. It is not difficult to show that the stress relation (4.7) implies both $\bar{B}^{(\infty)}_{ijpq} = \bar{B}^{(\infty)}_{ijpq}$ and $\bar{B}^{(0)}_{ijpq} = \bar{B}^{(0)}_{pqij}$.

** We first announced this result at a meeting of the Society of Rheology in Pittsburgh on October 26, 1964.

*** We may observe that since, under the hypotheses of Theorem 5.1, $\bar{\boldsymbol{Q}}(\boldsymbol{n})$ and $\hat{\boldsymbol{Q}}(\boldsymbol{n})$ are symmetric for every $\boldsymbol{n}$, $\bar{\boldsymbol{Q}}(\boldsymbol{n})$ and $\hat{\boldsymbol{Q}}(\boldsymbol{n})$ are determined by the quadrics, which in conformity with current practice in acoustics and optics may be called *Fresnel ellipsoids* or *ellipsoids of polarization*. Of course, such a quadric will not be a true ellipsoid unless the corresponding acoustic tensor is assumed to be positive definite, an assumption which may not hold in general.

we can write (5.5) in the form

$$D_{F_{jm}} \hat{\mathfrak{S}}_{il}(\boldsymbol{F}^t, \eta^t) = D_{F_{jm}} \overline{\mathfrak{S}}_{il}(\boldsymbol{F}^t, \vartheta^t) + \frac{\vartheta}{\varkappa} [D_\vartheta \overline{\mathfrak{S}}_{il}(\boldsymbol{F}^t, \vartheta^t)] [D_\vartheta \overline{\mathfrak{S}}_{jm}(\boldsymbol{F}^t, \vartheta^t)]. \quad (5.7)$$

Multiplying (5.7) by $F_{pm} F_{ql} n_p n_q$ and using (3.9) and (3.17), we arrive at the formula*

$$\hat{Q}_{ij}(\boldsymbol{n}) = \overline{Q}_{ij}(\boldsymbol{n}) + \frac{\vartheta}{\varkappa} P_i(\boldsymbol{n}) P_j(\boldsymbol{n}), \quad (5.8)$$

where

$$P_i(\boldsymbol{n}) = D_\vartheta \overline{\mathfrak{S}}_{il}(\boldsymbol{F}^t, \vartheta^t) F_{ql} n_q. \quad (5.9)$$

This proves

Theorem 5.2. *Let the functional transformation* (4.15) *be invertible, and suppose that the stress relation* (4.17) *and the temperature relation* (4.18) *hold. Then, the instantaneous homothermal and adiabatic acoustic tensors* $\overline{\boldsymbol{Q}}(\boldsymbol{n})$ *and* $\hat{\boldsymbol{Q}}(\boldsymbol{n})$ *are related by the formula* (5.8), *with* $\varkappa$ *given by* (4.22) *and* $P_i(\boldsymbol{n})$ *by* (5.9).

Let us assume that $\varkappa > 0$. Of course, the tensor with components $P_i(\boldsymbol{n}) P_j(\boldsymbol{n})$ is, by definition, positive semi-definite. Hence (5.8) implies that if $\overline{\boldsymbol{Q}}(\boldsymbol{n})$ is positive definite, then so also is $\hat{\boldsymbol{Q}}(\boldsymbol{n})$. Combining this observation with Theorems 3.1, 3.2, 4.1, 4.2, and 5.2, we justify

Remark 5.1. Suppose that the hypothesis of Theorem 4.1 holds, that the functional transformation (4.15) is smoothly invertible, and that $\varkappa$ is positive. Then, whenever $\boldsymbol{F}^t, \vartheta^t$ and $\boldsymbol{n}$ are such that all homothermal wave speeds are real, so also are all homentropic wave speeds.

For a perfectly elastic material (with a stored energy function) the assertion that

$$\boldsymbol{C} = \boldsymbol{C}^T \neq \boldsymbol{0} \Rightarrow \overline{B}^{(\infty)}_{ijpq} C_{ij} C_{pq} > 0, \quad (5.10)$$

is called the C-N$^+$ *condition***. In [4], TRUESDELL gives an argument (attributing it to NOLL and ERICKSEN) which shows that for perfectly elastic materials the C-N$^+$ condition implies that at least one wave speed is real. That argument may be generalized to waves in general materials obeying the principle of fading memory, provided we restrict attention to waves entering a region at rest.

Remark 5.2. Consider an acceleration wave entering a general material with memory which obeys the hypothesis of Theorem 4.1, and suppose that at the moment of arrival of the wave the histories $\boldsymbol{F}^t, \vartheta^t$ of the material are the constant histories $\boldsymbol{F}^\dagger, \vartheta^\dagger$ of (4.24). If the material obeys the condition (5.10) with $\overline{B}^{(\infty)}_{ijpq}$ defined by (4.25a) and (4.27a), then $\overline{\boldsymbol{Q}}(\boldsymbol{n})$ has at least one positive proper number, *i.e.*, for every value of $\boldsymbol{n}$, the homothermal propagation conditions (3.14) yield at least one real speed.

---

* For elastic materials (5.8) was proven by DUHEM; *cf.* TRUESDELL [4], § 13. We gave the one-dimensional analogue of (5.8), for materials with memory, in Remark 9.1 of III.

** Articles by TRUESDELL & TOUPIN [14] and TOUPIN & BERNSTEIN [15] explain the relation of (5.10) to the convexity inequality originally proposed by COLEMAN & NOLL in [16]. We use the terminology introduced in [14]. See also TRUESDELL [4] and COLEMAN & NOLL [17]. Here, when we say that a material is *perfectly elastic*, we mean not only that $\boldsymbol{S} = \overline{\boldsymbol{S}}(F, \vartheta)$ but also that $\overline{S}_{ij}(\boldsymbol{F}, \vartheta) = \frac{\partial}{\partial F_{ij}} \psi(\boldsymbol{F}, \vartheta)$, where $\boldsymbol{S} = \boldsymbol{S}(t)$, $\boldsymbol{F} = \boldsymbol{F}(t)$, and $\vartheta = \vartheta(t)$. Hence for such materials $\overline{\boldsymbol{Q}}(\boldsymbol{n}) = \overline{\boldsymbol{Q}}^T(\boldsymbol{n})$ and $\overline{B}^{(\infty)}_{ijpq} = \overline{B}^{(0)}_{ijpq} = \overline{B}^{(\infty)}_{pqij}$.

Proof. Since we assume that $\boldsymbol{F}^t = \boldsymbol{F}^\dagger$ and $\vartheta^t = \vartheta^\dagger$, it follows from (3.9) and (4.29a) that

$$\overline{Q}_{ij}(\boldsymbol{n}) = \overline{B}^{(0)}_{ipjq} n_p n_q, \tag{5.11}$$

and, since we also assume that the hypothesis of Theorem 4.1 is obeyed, we can use (4.32), which when combined with (5.10) yields

$$\boldsymbol{C} = \boldsymbol{C}^T \neq \boldsymbol{0} \Rightarrow \overline{B}^{(0)}_{ijpq} C_{ij} C_{pq} > 0. \tag{5.12}$$

Putting $C_{kl} = n_k n_l$ in (5.12), we obtain

$$\overline{B}^{(0)}_{ijpq} n_i n_j n_p n_q = Q_{ij}(\boldsymbol{n})\, n_i n_j > 0. \tag{5.13}$$

Now the proper numbers of $\overline{\boldsymbol{Q}}(\boldsymbol{n})$ are the extrema of the bounded continuous function

$$f(\boldsymbol{l}) = \overline{Q}_{ij}(\boldsymbol{n})\, l_i l_j \tag{5.14}$$

over the set of all unit vectors $\boldsymbol{l}$. Since, by (5.13), f has at least one positive value (the value at $\boldsymbol{l} = \boldsymbol{n}$), at least one extremum of $f(\boldsymbol{l})$ is positive. Hence $\overline{\boldsymbol{Q}}(\boldsymbol{n})$ has at least one positive proper number; q.e.d.

An analogous argument proves

Remark 5.3. Consider an acceleration wave entering at time t a general material with memory obeying the hypothesis of Theorem 4.1, and suppose that $\boldsymbol{F}^t = \boldsymbol{F}^\dagger$ and $\eta^t = \eta^\dagger$. If the material is such that

$$\boldsymbol{C} = \boldsymbol{C}^T \neq \boldsymbol{0} \Rightarrow \hat{B}^{(\infty)}_{ijpq} C_{ij} C_{pq} > 0$$

with $\hat{B}^{(\infty)}_{ijpq}$ defined by (4.25b) and (4.27b), then $\hat{\boldsymbol{Q}}(\boldsymbol{n})$ has at least one proper number; *i.e.*, for every value of $\boldsymbol{n}$, the homentropic propagation condition (3.18) yields at least one real speed.

6. Waves in Definite Conductors and Non-Conductors

We now investigate the physical circumstances in which acceleration waves are homothermal or homentropic.

A material is called a *definite conductor of heat* if the *heat conduction tensor* $\boldsymbol{K} = \boldsymbol{K}(t)$, defined by

$$\boldsymbol{K}_{ij} = -\frac{\partial}{\partial g_i} \bar{\mathfrak{q}}_j(\boldsymbol{F}^t, \vartheta^t, \boldsymbol{g}), \tag{6.1}$$

is always positive-definite.* A material is a *non-conductor* if $\bar{\mathfrak{q}}$ is the constant functional with value zero; *i.e.*, if

$$\boldsymbol{q} \equiv \boldsymbol{0}. \tag{6.2}$$

Let us first consider an acceleration wave in a definite conductor of heat. We assume that in addition to A-1 and A-1′, the following condition holds.

A-3:** The mappings $t \to \boldsymbol{F}^t_r(\boldsymbol{X}, \cdot)$ and $t \to \vartheta^t_r(\boldsymbol{X}, \cdot)$, with values in the function spaces $\mathcal{H}$ and $\mathscr{H}$, have derivatives $\dot{\boldsymbol{F}}^t_r(\boldsymbol{X}, \cdot)$, $\dot{\vartheta}^t_r(\boldsymbol{X}, \cdot)$ that are continuous in $\boldsymbol{X}$ and t for all $\boldsymbol{X}$ and t.

We continue to assume that the body force $\boldsymbol{b}$ and the heat supply r are continuous functions of $\boldsymbol{X}$ and t. As a consequence of balance of energy (4.5)

---

* In the case of Fourier's law, $\boldsymbol{q} = -\boldsymbol{K}\boldsymbol{g}$, the concept of a definite conductor is familiar.

** If the norms on $\mathcal{H}$ and $\mathscr{H}$ are of the fading memory type (1.4), and if the wave is entering a material for which $\boldsymbol{F}$ and ϑ have been constant in space and time, then A-3 is implied by A-1 and A-1′.

and the assumed smoothness of the functionals in (4.3), we have the familiar relation

$$[\boldsymbol{q}]\cdot\boldsymbol{n}=0. \tag{6.3}$$

Equation (6.3), when combined with the constitutive equation (4.3d), implies that

$$\boldsymbol{n}\cdot\left[\bar{\mathfrak{q}}(\boldsymbol{F}^t,\vartheta^t,\boldsymbol{g}+[\boldsymbol{g}])-\bar{\mathfrak{q}}(\boldsymbol{F}^t,\vartheta^t,\boldsymbol{g})\right]=0 \tag{6.4}$$

with $\boldsymbol{F}^t$ and ϑ^t histories at the wave and $\boldsymbol{g}$ the temperature gradient just ahead of the wave. Since $[\vartheta]=0$, it follows from a theorem of MAXWELL* that

$$[\boldsymbol{g}]=\alpha\,\boldsymbol{n}, \tag{6.5}$$

where α is a scalar. Therefore, (6.4) may be written

$$\boldsymbol{n}\cdot\left[\bar{\mathfrak{q}}(\boldsymbol{F}^t,\vartheta^t,\boldsymbol{g}+\alpha\,\boldsymbol{n})-\bar{\mathfrak{q}}(\boldsymbol{F}^t,\vartheta^t,\boldsymbol{g})\right]=0. \tag{6.6}$$

But, since the material is a definite conductor,

$$\frac{d}{d\alpha}\,\boldsymbol{n}\cdot\bar{\mathfrak{q}}(\boldsymbol{F}^t,\vartheta^t,\boldsymbol{g}+\alpha\,\boldsymbol{n})=\frac{\partial}{\partial g_i}\,\bar{\mathfrak{q}}_j(\boldsymbol{F}^t,\vartheta^t,\boldsymbol{g}+\alpha\,\boldsymbol{n})\,n_i\,n_j<0. \tag{6.7}$$

It follows from (6.7) that (6.6) can hold only if $\alpha=0$, *i.e.*, only if

$$[\boldsymbol{g}]\equiv[\nabla_{\boldsymbol{x}}\vartheta]=\boldsymbol{0}. \tag{6.8}$$

Since $[\vartheta]=0$, we have** $[\dot{\vartheta}]=-V[\nabla_{\boldsymbol{x}}\vartheta]\cdot\boldsymbol{n}=0$, and, by (4.2), $[\nabla_{\boldsymbol{X}}\vartheta]=\boldsymbol{F}^T[\nabla_{\boldsymbol{x}}\vartheta]=0$. This proves***

Theorem 6.1. *It is a consequence of the law of balance of energy that every acceleration wave in a definite conductor of heat is homothermal.*

Let us consider an acceleration wave obeying the hypotheses A-1, A-1′, A-2, A-2′, and A-3. On either side of the wave we may differentiate (4.3a) with respect to time:

$$\begin{aligned}\dot{\psi}&=\frac{d}{dt}\,\bar{\mathfrak{p}}(\boldsymbol{F}^t,\vartheta^t)\\&=D_{\boldsymbol{F}}\bar{\mathfrak{p}}(\boldsymbol{F}^t,\vartheta^t|\dot{\boldsymbol{F}})+\delta_{\boldsymbol{F}^t_r}\bar{\mathfrak{p}}(\boldsymbol{F}^t,\vartheta^t|\dot{\boldsymbol{F}}^t_r)+D_{\vartheta}\bar{\mathfrak{p}}(\boldsymbol{F}^t,\vartheta^t|\dot{\vartheta})+\delta_{\vartheta^t_r}\bar{\mathfrak{p}}(\boldsymbol{F}^t,\vartheta^t|\dot{\vartheta}^t_r).\end{aligned} \tag{6.9}$$

Here $\dot{\boldsymbol{F}}$ and $\dot{\vartheta}$ are present values, $\dot{\boldsymbol{F}}^t_r$ is the derivative of the mapping $t\to\boldsymbol{F}^t_r(\boldsymbol{X},\cdot)$ with values in $\mathcal{H}$, and $\dot{\vartheta}^t_r$ is the derivative of the mapping $t\to\vartheta^t_r(\boldsymbol{X},\cdot)$ with values in $\mathscr{H}$. By our hypothesis A-3, $\dot{\boldsymbol{F}}^t_r(\boldsymbol{X},\cdot)$ and $\dot{\vartheta}^t_r(\boldsymbol{X},\cdot)$ are continuous across the wave. Hence, the continuity of $\delta_{\boldsymbol{F}^t_r}\bar{\mathfrak{p}}$ and $\delta_{\vartheta^t_r}\bar{\mathfrak{p}}$ in all their variables insures that the jumps in the second and fourth terms on the right-hand side of (6.9) are zero, and, since

$$D_{\boldsymbol{F}}\bar{\mathfrak{p}}(\boldsymbol{F}^t,\vartheta^t|\dot{\boldsymbol{F}})=D_{F_{ij}}\bar{\mathfrak{p}}(\boldsymbol{F}^t,\vartheta^t)\,\dot{F}_{ij}, \tag{6.10a}$$

$$D_{\vartheta}\bar{\mathfrak{p}}(\boldsymbol{F}^t,\vartheta^t|\dot{\vartheta})=D_{\vartheta}\bar{\mathfrak{p}}(\boldsymbol{F}^t,\vartheta^t)\,\dot{\vartheta}, \tag{6.10b}$$

* *Vid.* [5], eq. (175.8).

** See eq. (180.4) of [5].

*** The argument we have just given is a generalization of one given by TRUESDELL [4] for materials obeying Fourier's Law and attributed by TRUESDELL to FOURIER; see also [5], § 296, and III, Theorem 4.2.

we have the relation

$$[\dot{\psi}] = D_{F_{ij}} \bar{\mathfrak{p}}(\boldsymbol{F}^t, \vartheta^t)[\dot{F}_{ij}] + D_\vartheta \bar{\mathfrak{p}}(\boldsymbol{F}^t, \vartheta^t)[\dot{\vartheta}]. \tag{6.11}$$

If we use the stress relation (4.7) and the entropy relation (4.8), we can write (6.11) in the form

$$[\dot{\psi}] = S_{ij}[\dot{F}_{ij}] - \eta[\dot{\vartheta}]. \tag{6.12}$$

Since A-1, A-1′, A-2, A-2′, and the smoothness of the functional $\bar{\mathfrak{h}}$ imply that

$$[\eta] = 0, \tag{6.13}$$

the equation (4.4) yields

$$[\dot{\varepsilon}] = [\dot{\psi}] + \vartheta[\dot{\eta}] + [\dot{\vartheta}]\eta, \tag{6.14}$$

and on combining this with (6.12) we obtain the useful relation*

$$[\dot{\varepsilon}] = S_{ij}[\dot{F}_{ij}] + \vartheta[\dot{\eta}]. \tag{6.15}$$

The continuity of $\boldsymbol{b}$ and r, the smoothness of the constitutive functionals in (4.3), our hypotheses on the wave, and the laws of balance of momentum and energy, imply that

$$[\dot{\varepsilon}] = S_{ij}[\dot{F}_{ij}] - \frac{1}{\varrho}\left[\frac{\partial q_i}{\partial x_i}\right], \tag{6.16}$$

and on substituting this into (6.15), we find that

$$\varrho\vartheta[\dot{\eta}] = -\left[\frac{\partial q_i}{\partial x_i}\right]. \tag{6.17}$$

We now assume that the acceleration wave is propagating in a non-conductor; *i.e.*, that (6.2) holds. Since $\vartheta > 0$, equation (6.17) reduces to $[\dot{\eta}] = 0$. Furthermore, it follows from (6.13) that** $[\dot{\eta}]^2 = V^2[\nabla_{\boldsymbol{x}}\eta]\cdot[\nabla_{\boldsymbol{x}}\eta]$. Hence, when $V \neq 0$, (6.2) and (6.17) imply both $[\dot{\eta}] = 0$ and $[\nabla_{\boldsymbol{X}}\eta] = \boldsymbol{F}^T[\nabla_{\boldsymbol{x}}\eta] = \boldsymbol{0}$. This proves

Theorem 6.2. *It is a consequence of the laws of balance of momentum and energy, the stress relation* (4.7), *and the entropy relation* (4.8) *that every acceleration wave with* $V \neq 0$ *in a non-conductor is homentropic.*

Our Theorems 6.1 and 6.2 generalize to materials with memory the second theorem of Duhem for elastic materials.

We thank Professor C. Truesdell for his criticism of an earlier draught of this manuscript.

The support of this research by the National Science Foundation, the Air Force Office of Scientific Research, the Advanced Research Projects Agency of the Department of Defense and the Office of Naval Research is gratefully acknowledged.

---

* Our present eqs. (6.12) and (6.15) generalize to three-dimensions eqs. (4.12) and (4.14) of III. Equation (6.15) can also be derived by using the constitutive eqs. (4.10) and by assuming that the hypotheses A-1, A-1′, A-2, A-2′, and A-3 are satisfied with ϑ replaced by η. In that derivation $\bar{\mathfrak{p}}$ is replaced by $\hat{\mathfrak{e}}$, (4.7) by (4.17), (4.8) by (4.18), and the argument which gave us (6.12) then gives (6.15) directly.

** We use eq. (180.6) of [5].

References

[1] COLEMAN, B. D., M. E. GURTIN, & I. HERRERA R., Arch. Rational Mech. Anal. **19**, 1—19 (1965).
[2] COLEMAN, B. D., & M. E. GURTIN, Arch. Rational Mech. Anal. **19**, 239—265 (1965).
[3] COLEMAN, B. D., & M. E. GURTIN, Arch. Rational Mech. Anal. **19**, 266—298 (1965).
[4] TRUESDELL, C., Arch. Rational Mech. Anal. **8**, 263—296 (1961).
[5] TRUESDELL, C., & R. A. TOUPIN, The Classical Field Theories. In the Encyclopedia of Physics, Vol. III/1, edited by S. FLÜGGE. Berlin-Göttingen-Heidelberg: Springer 1960.
[6] ERICKSEN, J. L., J. Rational Mech. Anal. **2**, 329—337 (1953).
[7] COLEMAN, B. D., Arch. Rational Mech. Anal. **17**, 1—46 (1964).
[8] COLEMAN, B. D., Arch. Rational Mech. Anal. **17**, 230—254 (1964).
[9] HADAMARD, J., Bull. Soc. Math. France **29**, 50—60 (1901).
[10] TRUESDELL, C., Internat. Sympos. Second-Order Effects, Haifa, 1962: London: Macmillan 1964, pp. 187—199.
[11] HERRERA, I., & M. E. GURTIN, Quart. Appl. Math. **22**, 360—364 (1965).
[12] VARLEY, E., Arch. Rational Mech. Anal. **19**, 215—225 (1965).
[13] COLEMAN, B. D., & W. NOLL, Arch. Rational Mech. Anal. **13**, 167—178 (1963).
[14] TRUESDELL, C., & R. TOUPIN, Arch. Rational Mech. Anal. **12**, 1—33 (1963).
[15] TOUPIN, R. A., & B. BERNSTEIN, J. Acoust. Soc. Amer. **33**, 216—225 (1961).
[16] COLEMAN, B. D., & W. NOLL, Arch. Rational Mech. Anal. **4**, 97—128 (1959).
[17] COLEMAN, B. D., & W. NOLL, Arch. Rational Mech. Anal. **15**, 87—111 (1964).
[18] GREEN, W. A., Arch. Rational Mech. Anal. **16**, 79—88 (1964).

Mellon Institute
Pittsburgh, Pennsylvania
and
Brown University
Providence, Rhode Island

(Received April 18, 1965)

General and Exact Theory of Waves in Finite Elastic Strain

C. TRUESDELL

Dedicated to HADAMARD *in admiration and affection on the sixtieth anniversary of his great treatise, Leçons sur la Propagation des Ondes*

Contents

1. Status of the problem. Summary

After the classical researches of CHRISTOFFEL, HUGONIOT, HADAMARD, and DUHEM on waves in elastic materials, it might seem that little remains to be learned. Such is not the case. As for most parts of mechanics, it has been necessary in the last decade to go over the matter again, not only so as to free the conceptual structure from lingering linearizing and to fix it more solidly in the common foundation of modern mechanics, but also so as to derive from it specific predictions satisfying modern needs for contact between theory and rationally conceived experiment. After reading the recent papers by TOUPIN & BERNSTEIN [1961, 1] and by HAYES & RIVLIN [1961, 2], I have seen that more can be learned than is there proved. In the present paper I follow TOUPIN & BERNSTEIN's approach to the general theory yet try to achieve the elegant and explicit directness of ERICKSEN's earlier treatment of isotropic incompressible materials [1953]. At the same time, all the results of HAYES & RIVLIN are obtained in shorter but more general form as immediate corollaries.

It should not be thought, however, that this paper serves only as a condensing summary. Rather, I consider the general theory of elasticity instead of its special case, which was the starting point of all previous authors and which will here be called *hyperelastic*, when the stresses are derivable from a stored energy. The more general elastic theory, though as yet but dimly known, is of some interest in itself: In statics, any simple material according to the definition of NOLL [1958] is an elastic material, though not necessarily a hyperelastic one, and also RIVLIN [1959, 2] [1960] has shown that for certain classes of deformations a fairly general kind of material is approximately elastic, though again not necessarily hyperelastic. These observations, while not particularly relevant to the problem of wave propagation, give the more general theory of elasticity some individual status, although there are various energy theorems standing in favor of the hyperelastic special case.

In a given elastic material subject to a given state of strain, a wave may propagate in a given direction only if its amplitude is parallel to an *acoustic axis;* corresponding to each axis is a uniquely determined absolute speed of propagation. In a hyperelastic material there always exists an orthogonal real triad of acoustic axes, though the corresponding speeds are not necessarily real. For a general elastic material, the possibilities are more various and less definite; in some cases there may be but one real acoustic axis, and if there are three, they need not be orthogonal. However, it turns out that *the assumption of isotropy alone*, regardless of whether or not a stored energy exists, suffices to derive some definite acoustical theorems in a generality greater than any heretofore known even for the hyperelastic case.

These theorems may be foreshadowed by physical reasoning. Waves travelling in the direction of a unit vector $\boldsymbol{n}$ in a general elastic material face a complex of three triads of generally distinct special directions: (1) the vectors $\boldsymbol{d}_A$ in terms of which the symmetries of the material may be specified, (2) the principal axes of stress, (3) the principal axes of strain in the deformed material. From these three sets of directions, as well as from the local rotation, the acoustic axes for the direction $\boldsymbol{n}$ are somehow determined. In an isotropic material, however, the vectors $\boldsymbol{d}_A$ are arbitrary, the principal axes of stress and strain coincide, and the local rotation is irrelevant. Thus the acoustic axes depend upon one orthogonal set of directions in the material and, of course, upon $\boldsymbol{n}$. When $\boldsymbol{n}$ is parallel to a principal axis of strain and stress, there remain but three directions, namely, the principal axes themselves, to influence the acoustic response. Thus we may expect, and indeed I shall prove, that *in an isotropic elastic material the acoustical axes for a wave travelling down a principal axis of stress and strain coincide with the principal axes.* In particular, *a wave travelling down a principal axis of stress and strain in an isotropic body is always either longitudinal or transverse.* Finally, for these waves I derive *explicit and exact formulae for the speeds of propagation.* Thus for isotropic elastic materials the presence or absence of a stored energy is of secondary importance in problems of wave propagation.

These same results may be motivated also by starting from the infinitesimal theory. In that theory, the material is virtually unstressed and unstrained as far as the passage of waves is concerned. The only preferred directions are those of the material itself, and these determine the acoustic axes for each direction of propagation. In the isotropic case, there are no preferred material directions, and the only direction which can influence the acoustic axes is that of the wave normal. Hence, as is well known, every wave is necessarily either normal or transverse; any transverse amplitude is possible; and all waves of any one of these two kinds have a common absolute speed of propagation. In a finitely strained body, in general, this simple result holds no longer[1]. Rather, on a wave travelling in a severely strained isotropic body the existence of principal axes of stress and strain must have much the same effect as does that of the axes of symmetry of an unstrained anisotropic material. Only when the wave itself travels down one of these axes is the total number of special directions reduced to three, and there is then no reason for the acoustic axes to be other than these. Thus, again, we are led to expect the results announced above.

But for one special case, namely, that of a material subject to hydrostatic pressure, simplicity comparable to that of the linearized theory remains in force for strains of any magnitude. In this case, I show that every wave is necessarily either transverse or longitudinal; that all waves of each kind have the same absolute speed; and that a simple identity connects the bulk modulus, the absolute speed of the longitudinal waves, and the absolute speed of the transverse waves. It follows that longitudinal waves always travel faster than transverse waves, and that if transverse waves are possible, the absolute speed of longitudinal waves is always greater than the speed of sound according to the hydrodynamical theory for the same law of compression.

Returning to arbitrary states of strain, I derive a number of universal formulae, valid for all possible forms of the stress-strain relations for an isotropic material, which connect the speeds of the 9 different kinds of waves that can travel along the principal axes of stress and strain. In particular, the speeds of any 2 of the 6 possible kinds of transverse waves determine explicitly those of the remaining 4. These formulae serve as *conditions of compatibility* which must be satisfied by measured values of the speeds if they are to be consistent with the theory of elasticity for isotropic materials. Supposing these conditions to be satisfied, I find simple explicit equations determining the form of the stress-strain relations directly from the speeds of the longitudinal waves along the three principal axes and from the speeds of the two kinds of transverse waves along any one of them.

From these results follows a differential criterion whereby it can be determined from the wave speeds as functions of the principal stretches whether or not an isotropic elastic material be hyperelastic.

Conditions under which the wave speeds are real are then investigated. Among other things, it is shown that the inequalities of BAKER & ERICKSEN are necessary and sufficient that all six kinds of transverse waves in isotropic materials have real speeds of propagation.

By expanding the exact formulae for the wave speeds in powers of the three principal extensions, I calculate the wave speeds according to the second-order theory of elasticity. The results of L. BRILLOUIN for hydrostatic pressure and

[1] HADAMARD [1903, pp. 274—277] determined the very special form the stored energy must have in order that every wave, whatever its direction and whatever the strain, be necessarily either longitudinal or transverse.

of TOUPIN & BERNSTEIN for bi-axial strain are included as special cases. Certain algebraic conditions must be satisfied by the 9 wave speeds in order that they be compatible with the second-order theory; the 4 second-order elasticities are then determined explicitly by any 2 transverse-wave speeds and any 2 longitudinal-wave speeds. I show also that a simple relation between one longitudinal-wave speed and one transverse-wave speed is *a necessary and sufficient condition that there be a stored energy* in the second-order theory. So far as I know, this result affords the first chance for a direct and clear experimental test of the existence of a stored energy.

Finally I present new and simple proofs of three theorems of DUHEM connecting the wave speeds in elastic materials conducting heat according to FOURIER'S law with those in elastic materials that do not conduct heat at all.

2. General theory of acceleration waves[1]

At a singular surface of second order with unit normal $\boldsymbol{n}$ the following geometrical and kinematical conditions of compatibility [CFT, Eqs. (190.1), (190.2), (190.5)] must be satisfied by the jumps $[x^k_{,\alpha\beta}]$ and $[\ddot{x}^k]$ in the second derivatives $x^k_{,\alpha\beta}$ of the deformation $x^k = x^k(X^\alpha, t)$ and by the jump of the acceleration $\ddot{x}^k$:

$$[x^k_{,\alpha\beta}] = a^k\, x^m_{,\alpha}\, x^p_{,\beta}\, n_m\, n_p, \qquad [\ddot{x}^k] = U^2 a^k, \tag{2.1}$$

where $\boldsymbol{a}$ is a vector which may be called the *amplitude* and where U is the *speed of propagation*. The corresponding dynamical condition [CFT, Eq. (205.5)] is

$$[t^{km}]\, n_m = 0, \tag{2.2}$$

where $\boldsymbol{t}$ is CAUCHY'S stress tensor.

To begin with, we take the constitutive equation of elasticity in fully general curvilinear co-ordinate systems at $\boldsymbol{X}$ and $\boldsymbol{x}$ in the form[2]

$$T_k^\alpha = \mathfrak{h}_k^\alpha(x^m_{,\beta}, \boldsymbol{d}_A), \tag{2.3}$$

where $\boldsymbol{T}$ is the first PIOLA-KIRCHHOFF stress tensor[3], where $\mathfrak{h}$ is an arbitrary function which is assumed to be as smooth as desired, and where the three vectors $\boldsymbol{d}_A$ serve to specify the symmetries of the material at the particle in question. The dynamical condition (2.2) is satisfied trivially by (2.3), since $x^m_{,\beta}$,

---

1 While this paper is self-contained, the reader is expected to know the principles of modern continuum mechanics; for ease of reference in case of doubt, by using the prefix CFT we cite the relevant equations or sections in TRUESDELL & TOUPIN'S *The Classical Field Theories*, with an appendix on *Invariants* by J. L. ERICKSEN, in FLÜGGE'S Handbuch der Physik, vol. 3/1, p. 225–793, 1960.

2 From these formulae it is clear that the method of FINZI [1942], which is akin to that we use in §7 for isotropic materials, is not general. In fact, FINZI expressly limited himself to infinitesimal deformations, although he allowed the stress-strain relations to be of arbitrary form.

3 *Cf.* CFT, §210. For the purposes of the present paper we may take $\boldsymbol{T}$ as defined formally by the relation

$$T^{k\alpha} = \frac{\tilde{\varrho}}{\varrho}\, t^{km}\, X^\alpha_{,m}, \tag{A}$$

where ϱ and $\tilde{\varrho}$ are the mass densities in the present configuration and in the reference configuration, respectively.

and hence also $\boldsymbol{t}$, is continuous at the singular surface. In terms of $\boldsymbol{T}$, CAUCHY'S first law of motion assumes the form [CFT, Eq. (210.8)]

$$T_{k\,;\alpha}^{\alpha} + \tilde{\varrho}\, f_k = \tilde{\varrho}\, \ddot{x}_k, \tag{2.4}$$

where $\boldsymbol{f}$ is the extrinsic force per unit mass, where $\tilde{\varrho}$ is the mass density in the reference configuration, and where the semicolon denotes the total covariant derivative [CFT, App., §20].

Let us set

$$A_{k\,m}{}^{\alpha\beta}(x^p{}_{,\gamma}, \boldsymbol{d}_A) \equiv \frac{\partial \mathfrak{h}_k^{\alpha}}{\partial x^m{}_{,\beta}}. \tag{2.5}$$

Substitution of (2.3) into (2.4) then yields equations of motion in the form

$$A_{k\,m}{}^{(\alpha\beta)}\, x^m{}_{;\alpha\beta} + \frac{\partial \mathfrak{h}_k^{\alpha}}{\partial d_A^m}\, d^m_{A;\alpha} + \tilde{\varrho}\, f_k = \tilde{\varrho}\, \ddot{x}_k. \tag{2.6}$$

Assuming that $A_{k\,m}{}^{(\alpha\beta)}$, $\partial \mathfrak{h}_k^{\alpha}/\partial d_A^m$, $d^m_{A;\alpha}$, $\tilde{\varrho}$, and f_k are continuous across the singular surface[4], we calculate the jump of (2.6), use (2.1), and so obtain the following *propagation condition*

$$Q_{k\,m}(\boldsymbol{n})\, a^m = \tilde{\varrho}\, U^2 a_k, \tag{2.7}$$

where

$$\begin{aligned} Q_{k\,m}(\boldsymbol{n}) &\equiv A_{k\,m}{}^{(\alpha\beta)}\, x^p{}_{,\alpha}\, x^q{}_{,\beta}\, n_p\, n_q, \\ &= \frac{1}{2}\left(\frac{\partial \mathfrak{h}_k^{\alpha}}{\partial x^m{}_{,\beta}} + \frac{\partial \mathfrak{h}_k^{\beta}}{\partial x^m{}_{,\alpha}}\right) x^p{}_{,\alpha}\, x^q{}_{,\beta}\, n_p\, n_q. \end{aligned} \tag{2.8}$$

The tensor $\boldsymbol{Q}(\boldsymbol{n})$ may be called the *acoustical tensor*[5] for the direction $\boldsymbol{n}$ in the elastic material with response function $\mathfrak{h}$, subject to the deformation gradient $x^m{}_{,\alpha}$. From (2.7) we read off the first main result, which may justly be called the ***Fresnel-Hadamard theorem*** in fully general form: *The amplitude of a wave travelling in the direction of* $\boldsymbol{n}$ *must be a right proper vector of the acoustical tensor* $\boldsymbol{Q}(\boldsymbol{n})$; *the speed of propagation* U *must be such that* $\tilde{\varrho}\, U^2$ *is the corresponding proper number; conversely, any real right proper vector of* $\boldsymbol{Q}(\boldsymbol{n})$ *is a possible amplitude, provided the corresponding proper number is real and not negative.*

The speeds may be determined from the roots of the real cubic equation

$$\det\left(Q^k{}_m - \tilde{\varrho}\, U^2\, \delta^k_m\right) = 0 \tag{2.9}$$

and hence may be exhibited explicitly as algebraic functions of the principal invariants of $\boldsymbol{Q}(\boldsymbol{n})$; however, if it is known that $\boldsymbol{a}$ is a possible amplitude, it is easier to determine the corresponding squared speed from the formula

$$\tilde{\varrho}\, U^2 = Q_{(k\,m)}\, a^k\, a^m / a^2, \tag{2.10}$$

which follows at once from (2.7).

---

[4] In any uniformly aeolotropic material $d^m_{A;\alpha}$ is in fact identically zero. However, at the interface of two dissimilar homogeneous materials, the above-stated conditions are not satisfied, for $\mathfrak{h}$ itself fails to be continuous.

[5] In the hyperelastic theory, where, as we shall see in §12, $\boldsymbol{Q}(\boldsymbol{n})$ is symmetric for every $\boldsymbol{n}$, the quadric of $\boldsymbol{Q}(\boldsymbol{n})$ is called *Fresnel's ellipsoid* or the *ellipsoid of polarisation*. Since $\boldsymbol{Q}(\boldsymbol{n})$ in the generality maintained here is not necessarily symmetric, it is not determined by its quadric, so no particular name is given to this latter.

We note that $\boldsymbol{Q}(-\boldsymbol{n})=\boldsymbol{Q}(\boldsymbol{n})$; that $\boldsymbol{a}$ is a possible amplitude if and only if $k\boldsymbol{a}$ is also a possible amplitude for every real k; and that U is a possible speed of propagation for a wave travelling in the direction of $\boldsymbol{n}$ with amplitude $\boldsymbol{a}$ if and only if $-U$ is. Beyond the obvious symmetries corresponding to these properties of $\boldsymbol{Q}(\boldsymbol{n})$, none other is to be expected in the general theory.

3. General theory of waves of higher order

For a singular surface of third order we have, among others, the following geometrical and kinematical conditions of compatibility [CFT, §191]:

$$\begin{aligned} [x^k_{,\alpha\beta\gamma}] &= a^k\, x^m_{,\alpha}\, x^p_{,\beta}\, x^q_{,\gamma}\, n_m\, n_p\, n_q\,, \\ [\ddot{x}^k_{,m}] &= U^2\, a^k\, n_m\,. \end{aligned} \tag{3.1}$$

Differentiation of (2.6) with respect to x^p yields

$$A_{k\,m}^{(\alpha\beta)}\, x^m_{;\alpha\beta\gamma}\, X^\gamma_{,p} + \frac{\partial A_{k\,m}^{(\alpha\beta)}}{\partial x^q_{,\delta}}\, x^q_{;\delta\gamma}\, X^\gamma_{,p}\, x^m_{;\alpha\beta} + \left(\frac{\partial \mathfrak{h}_k^\alpha}{\partial d^m_A}\, d^m_{A;\alpha}\right)_{;\gamma} X^\gamma_{,p} + \tilde{\varrho}\, f_{k,p} = \tilde{\varrho}\, \ddot{x}_{k,p}\,. \tag{3.2}$$

If we assume that $f_{k,p}$, d^m_A, $d^m_{A;\alpha}$, and $d^m_{A;\alpha\gamma}$ are continuous across the singular surface, by taking the jump of (3.2) and employing (3.1) we conclude that

$$A_{k\,m}^{(\alpha\beta)}\, X^\gamma_{,p}\, a^m\, x^q_{,\alpha}\, x^r_{,\beta}\, x^s_{,\gamma}\, n_q\, n_r\, n_s = \tilde{\varrho}\, U^2\, a_k\, n_p\,, \tag{3.3}$$

a formula which reduces at once to (2.7). Similar analysis may be applied to any singular surface of higher order. If we are willing to presume that the vectors $\boldsymbol{d}_A$, the field of body force $\boldsymbol{f}$, and the functions $\mathfrak{h}_k^\alpha$ in the constitutive equation (2.3) are arbitrarily many times continuously differentiable, we may therefore assert the following ***first theorem of equivalence***[1]: *$\boldsymbol{a}$ and U are a possible amplitude and speed of propagation for a wave of order p travelling in the direction $\boldsymbol{n}$ if and only if they are likewise possible for a wave order q, where p and q are any two integers greater than* 1. Thus by considering waves of second order we exhaust the whole theory of singular surfaces other than shock waves and first or zero^th order vortex sheets {in so far as it is contained in the propagation condition}.

4. General theory of plane infinitesimal progressive waves in a material subject to homogeneous strain

A displacement of the form

$$\boldsymbol{u} = \boldsymbol{a}\sin(\boldsymbol{n}\cdot\boldsymbol{P} - U t)\,, \tag{4.1}$$

where $\boldsymbol{a}$ and $\boldsymbol{n}$ are constant vectors and $\boldsymbol{P}$ is the position vector, is said to represent a *sinusoidal plane progressive wave* of amplitude $\boldsymbol{a}$, propagating in the direction $\boldsymbol{n}$ at speed U. In general, as is clear from (2.6), a displacement of this kind is not possible in an elastic material. There is, however, an important special case in which (2.6) may be replaced, approximately, by a linear system with constant coefficients, namely, when $\boldsymbol{u}$ is an infinitesimal displacement superimposed upon a given state of homogeneous strain. While many authors

[1] While I cannot cite a previous statement of this theorem, it is only to be expected in view of known theorems for fluids and of ERICKSEN's corresponding theorem for isotropic incompressible hyperelastic materials [1953, 1, §6].

during the past seventy years have derived various forms of the appropriate differential equations of motion, the simplest and most general treatment is that of TOUPIN & BERNSTEIN [1961, 1, §2], which we shall here employ. Letting the co-ordinates X^K refer to the homogeneously strained initial state, in which the density is ϱ_0, we may write the differential equations for the additional infinitesimal displacement $\boldsymbol{u}$ in the form

$$\mathsf{K}_{MQ}{}^{(PR)}\, u^Q{}_{,PR} = \varrho_0 \ddot{u}_M, \tag{4.2}$$

where we have assumed that $d_{\Lambda;P}=0$ and that no extrinsic force is acting[1], and where

$$\mathsf{K}_{MQ}{}^{PR} = \frac{\varrho_0}{\tilde{\varrho}}\, g^m_M\, g^q_Q\, \mathsf{X}^P{}_{,\alpha}\, \mathsf{X}^R{}_{,\beta}\, A_{m\,q}{}^{\alpha\beta}, \tag{4.3}$$

g^r_R being the shifter[2] effecting finite parallel displacement from $\mathbf{X}$ to $\boldsymbol{x}$, and the argument of $A_{m\,q}{}^{\alpha\beta}$ being the gradient of the deformation from the natural state, with co-ordinates X^α, to the initial state, with co-ordinates X^K. So as to derive (4.2) it is assumed that the deformation from $\boldsymbol{X}$ to $\mathbf{X}$ is *homogeneous;* there is no reason to expect, in general, that (4.2) will hold otherwise[3].

Substitution of (4.1) into (4.2) leads to the condition

$$\mathsf{K}_{MQ}{}^{(PR)}\, n_P\, n_R\, a^Q = \varrho_0\, U^2\, a_M. \tag{4.4}$$

Since the superimposed periodic displacement $\boldsymbol{u}$ is assumed to correspond to an infinitesimal deformation, $\mathsf{X}^M{}_{,\alpha} \approx g^M_k\, x^k{}_{,\alpha}$, and hence by (4.3) and (2.8) we see that

$$\begin{aligned} \frac{\tilde{\varrho}}{\varrho}\, \mathsf{K}_{MQ}{}^{(PR)}\, n_P\, n_R &= g^m_M\, g^q_Q\, A_{m\,q}{}^{(\alpha\beta)}\, n_P\, \mathsf{X}^P{}_{,\alpha}\, n_R\, \mathsf{X}^R{}_{,\beta}, \\ &= g^m_M\, g^q_Q\, A_{m\,q}{}^{(\alpha\beta)}\, x^p{}_{,\alpha}\, x^r{}_{,\beta}\, g^P_p\, n_P\, g^R_r\, n_R, \\ &= g^m_M\, g^q_Q\, Q_{m\,q}. \end{aligned} \tag{4.5}$$

In virtue of this result, (4.4) has just the same form as does (2.7) shifted from $\boldsymbol{x}$ to $\mathbf{X}$. Since such a displacement is negligible in the present case, we obtain the ***second theorem of equivalence***[4]: *at a point in an elastic material where the deformation gradient is $x^k{}_{,\alpha}$, the amplitude $\boldsymbol{a}$ and the speed U satisfy the propagation condition for a singular surface of order 2 or more with unit normal $\boldsymbol{n}$ if and only if a plane infinitesimal sinusoidal wave of amplitude $\boldsymbol{a}$ may propagate in the direction of $\boldsymbol{n}$ at speed U through a specimen of the material subject to a homogeneous strain in which the deformation gradient is $x^k{}_{,\alpha}$.*

[1] Since the differential equations are now linear, extrinsic force may be taken into account by the addition of a particular integral. If the extrinsic force is independent of time, so is the corresponding integral; therefore time-independent forces have no effect on the propagation of infinitesimal waves.

[2] For explanation, see CFT, App., §16. The reader content with a single Cartesian system may set $g^k_K = \delta^k_K$. The devotees of convected co-ordinates must resort to more elaborate expedients.

[3] For arbitrary gradients $\mathsf{X}^K{}_{,\alpha}$ the differential equations for $\boldsymbol{u}$ are of the form

$$(\mathsf{K}_{MQ}{}^{PR}\, u^Q{}_{,R})_{,P} = \varrho_0\, \ddot{u}_M \tag{B}$$

as shown by TOUPIN & BERNSTEIN [1961, 1, Eq. (2.19)].

[4] This theorem generalizes a number of classical results in more special theories; *cf.* CFT, §194A.

5. Contrast and comparison of different types of waves

In virtue of the two theorems of equivalence, we are justified in interpreting any result in the theory of wave propagation either in terms of singular surfaces or in terms of plane infinitesimal oscillations, as we please. The mathematical status of these interpretations are not equal, however. The theory of infinitesimal oscillations results from "approximations" of as yet unestablished validity; also, it refers only to an underlying homogeneous strain and applies, at least as so far developed, only to plane waves. The speeds of propagation U are always constant. The theory of singular surfaces is exact; it applies equally to all amplitudes, all underlying states of strain, and all shapes of waves[1]. The speeds of propagation depend upon the existing strain and thus may vary with place and time. In this sense, the interpretation of results of the theory in terms of acceleration waves is both general and free of objection.

On the other hand, the relevance of the theory of singular surfaces is uncertain. That theory refers, strictly, only to necessary conditions for solutions. If there is a solution in which a singular surface with unit normal $\boldsymbol{n}$ carries a jump of amplitude $\boldsymbol{a}$, then (2.7) must be satisfied at each point, but a solution to (2.7) for $\boldsymbol{a}$ and $\boldsymbol{n}$ varying smoothly over a set of points does not generally yield a singular surface. First, it is necessary that at each time the unit normals $\boldsymbol{n}$ correspond to surface elements that sweep out a surface. Second, as is clear from known examples in the classical linearized theory, the dynamical equations (2.6) impose conditions on the motion of the singular surface and on the variation of $\boldsymbol{a}$ in time. The only deformation known to be possible for all uniformly aeolotropic materials is a state of homogeneous strain. Since $x^k{}_{,\alpha\beta}=0$ in any such state, it follows that $[x^k{}_{,\alpha\beta}]=0$, so that no non-trivial singular surface of second or higher order can travel through such a state. Thus, at present, no solution corresponding to any propagating singular surface is known.

These facts raise the disquieting thought that the whole theory of singular surfaces may turn out to be an empty formalism. I doubt, however, that this danger is real. It is clear that solutions consisting of two different homogeneous strains connected by a plane shock wave exist. As the initial strength of this shock wave is taken smaller and smaller, we expect that the relation between its amplitude and its speed will tend to satisfy the propagation condition for waves of second order. Analysis toward this end has been given by JOUGUET [1920, 1—4] [1921], and doubtless a general theorem of equivalence corresponding to more modern views on the thermodynamics of deformation could be established. In this way should result a universal status for plane waves of second order, and possibly also for waves of certain curved forms, as limiting cases of weak shocks. A general theory of shock waves in elastic materials is reserved for later study.

In this paper I present only a fairly complete *analysis of the propagation condition* (2.7), leaving for the future the search for corresponding solutions. So as to avoid lengthly wording, the necessary and sufficient criteria given in the following theorems refer *only to the propagation condition*, with no implication that singular surfaces corresponding to the results actually exist.

[1] Another difference may be seen in the effects of extrinsic forces: A *continuous* extrinsic force has no effect on the propagation of singular surfaces, while a *time-independent* extrinsic force has no effect on infinitesimal oscillations.

The interpretation in terms of infinitesimal oscillations has one advantage in that for it the non-positive proper numbers of $\boldsymbol{Q}(\boldsymbol{n})$ may be interpreted. For if $U^2<0$, then $U=\pm i\alpha$, where $\alpha>0$; from a linear combination of complex solutions of the type (4.1) we may construct a real solution of the form

$$\boldsymbol{u}=\boldsymbol{a}\, e^{\alpha t}\sin(\boldsymbol{n}\cdot\boldsymbol{P}), \tag{5.1}$$

and the magnitude of such a solution increases beyond all bounds as $t\to\infty$. Thus, if the linearized theory may be trusted, there are certain disturbances which, however small they may be initially, in time grow large enough to destroy the presumed underlying homogeneous strain. Thus that homogeneous strain is *unstable* in the sense of infinitesimal oscillations[2]. Complex roots U^2 occur in conjugate pairs and lead to conjugate proper vectors, so that again a real solution of exponentially increasing magnitude may be formed.

We should not let this fact induce us to require the constitutive equation (2.3) to be such that $\boldsymbol{Q}(\boldsymbol{n})$ have only positive proper numbers for all directions $\boldsymbol{n}$ and all states of strain. Rather, it is possible that physical instabilities such as buckling observed for strains of a certain magnitude result from the inability of a material when strained to that degree to allow the passage of tiny waves, especially transverse waves, without increasing their magnitude so much as to carry the body over into a state of small oscillation about a different configuration.

6. Consequences of the principle of material indifference

The theory studied so far is based on the constitutive equation (2.3) and hence is more general than elasticity, since it has not been required to satisfy the principle of material indifference [CFT, §239ζ]. It includes, for example, MacCullagh's theory of the rotationally elastic aether[1] [CFT, §302] and hence may be applied also in electromagnetic theory. As one of the many equivalent forms in which the effect of imposing the principle of material indifference may be put, we select the following specialization of (2.3) [CFT, Eq. (303.2)]:

$$T^{k\alpha}=x^k_{,\beta}\,\mathfrak{t}^{\beta\alpha}(\boldsymbol{C}) \tag{6.1}$$

where

$$\mathfrak{t}^{\beta\alpha}=\mathfrak{t}^{\alpha\beta}, \tag{6.2}$$

and where $\boldsymbol{C}$ is Green's deformation tensor [CFT, §26]. Hence from (2.3)

$$\mathfrak{h}_k^{\alpha}=g_{km}\,x^m_{,\beta}\,\mathfrak{t}^{\alpha\beta}. \tag{6.3}$$

---

[2] Hayes & Rivlin [1961, 2] call such a homogeneous strain "inherently unstable".

[1] Nowhere in the foregoing sections has Cauchy's second law of motion, $t^{km}=t^{mk}$, been imposed. In terms of the response function $\mathfrak{h}$, it is equivalent to the additional condition

$$\mathfrak{h}^{k\alpha}\,x^m_{,\alpha}=\mathfrak{h}^{m\alpha}\,x^k_{,\alpha}, \tag{B}$$

as follows from (2.3) and Eq. (A) in footnote 3, p. 266. In MacCullagh's theory, on the other hand, $t^{km}=-t^{mk}$. In the present ordering of ideas, Cauchy's second law is imposed here along with the principle of material indifference, the two together being expressed by (6.1) and (6.2).

Since $C_{\gamma\delta}=C_{\delta\gamma}$, there is no loss in generality if we suppose $\mathfrak{t}^{\alpha\beta}$ to be formally symmetrized in the components $C_{\gamma\delta}$. Thus we may put (2.5) into the form

$$\begin{aligned} A_{k m}{}^{\alpha\beta} &= g_{k m}\,\mathfrak{t}^{\alpha\beta} + g_{k q}\, x^q{}_{,\lambda} \frac{\partial \mathfrak{t}^{\alpha\lambda}}{\partial C_{\mu\nu}} \frac{\partial C_{\mu\nu}}{\partial x^m{}_{,\beta}}, \\ &= g_{k m}\,\mathfrak{t}^{\alpha\beta} + 2 g_{k p}\, g_{m q}\, x^p{}_{,\gamma}\, x^q{}_{,\delta} \frac{\partial \mathfrak{t}^{\alpha\gamma}}{\partial C_{\beta\delta}}. \end{aligned} \tag{6.4}$$

Let us set

$$C^{k m p q} \equiv 2 \frac{\varrho}{\tilde{\varrho}}\, x^k{}_{,\alpha}\, x^m{}_{,\beta}\, x^p{}_{,\gamma}\, x^q{}_{,\delta} \frac{\partial \mathfrak{t}^{\alpha\beta}}{\partial C_{\gamma\delta}}. \tag{6.5}$$

Then

$$C^{k m p q} = C^{m k p q} = C^{k m q p}; \tag{6.6}$$

these relations of symmetry reflect the principle of material indifference.

If we now set

$$B^{k m p q} \equiv \frac{\varrho}{\tilde{\varrho}}\, x^m{}_{,\alpha}\, x^q{}_{,\beta}\, g^{k r}\, g^{p s}\, A_{r s}{}^{\alpha\beta}, \tag{6.7}$$

so that

$$A_{k m}{}^{\alpha\beta} = \frac{\tilde{\varrho}}{\varrho}\, g_{k r}\, g_{m s}\, X^\alpha{}_{,p}\, X^\beta{}_{,q}\, B^{r p s q}, \tag{6.8}$$

then from (6.4) and (6.5) we see that[2]

$$B^{k m p q} = g^{k p}\, t^{m q} + C^{k m p q}. \tag{6.9}$$

Hence by (6.6) it follows that

$$\begin{aligned} B^{k m p q} - B^{m k p q} &= g^{k p}\, t^{m q} - g^{m p}\, t^{k q}, \\ B^{k m p q} - B^{k m q p} &= g^{k p}\, t^{m q} - g^{k q}\, t^{m p}, \\ B^{k m p q} - B^{m k q p} &= g^{k p}\, t^{m q} - g^{m q}\, t^{k p}. \end{aligned} \tag{6.10}$$

These identities, again, reflect the principle of material indifference.

In the classical infinitesimal theory, the tensors $A_{k m}{}^{(\alpha\beta)}$, $B^{k m p q}$, and $C^{k m p q}$ all reduce to the tensor of linear elasticities.

Substituting (6.8) into (2.8) yields an expression for the acoustical tensor in terms of $B^{k m p q}$:

$$\begin{aligned} Q^{k m} &= \frac{\tilde{\varrho}}{\varrho}\, X^{(\alpha}{}_{,u}\, X^{\beta)}{}_{,v}\, x^p{}_{,\alpha}\, x^q{}_{,\beta}\, B^{k u m v}\, n_p\, n_q, \\ &= \frac{1}{2} \frac{\tilde{\varrho}}{\varrho} \cdot (\delta^p_u\, \delta^q_v + \delta^q_u\, \delta^p_v)\, B^{k u m v}\, n_p\, n_q, \\ &= \frac{\tilde{\varrho}}{\varrho}\, B^{k p m q}\, n_p\, n_q. \end{aligned} \tag{6.11}$$

Another consequence of the principle of material indifference is the fact that the *speeds of propagation are determined by the strain alone, independently of the local rotation.* This is most easily seen by demonstrating an alternative form of the propagation condition (2.7), expressed in terms of material rather than spatial co-ordinates. Set

$$S^\alpha_\beta \equiv X^\alpha{}_{,k}\, Q^k{}_m\, x^m{}_{,\beta}, \qquad \nu_\alpha \equiv x^k{}_{,\alpha}\, n_k. \tag{6.12}$$

---

[2] This result was given by TOUPIN & BERNSTEIN [1961, 1, Eq. (2.23)] for the case of hyperelastic materials.

Then from (6.11) and (6.9) we see that[3]

$$\begin{aligned}
S^{\alpha}_{\beta} &= \frac{\tilde{\varrho}}{\varrho} X^{\alpha}_{,k} B^{kpsq} g_{ms} n_p n_q x^m_{,\beta}, \\
&= \frac{\tilde{\varrho}}{\varrho} X^{\alpha}_{,k} x^m_{,\beta} [\delta^k_m t^{pq} + g_{ms} C^{kpsq}] n_p n_q, \\
&= \left[\frac{\tilde{\varrho}}{\varrho} \delta^{\alpha}_{\beta} t^{pq} + 2 C_{\beta\lambda} x^p_{,\delta} x^q_{,\mu} \frac{\partial t^{\alpha\delta}}{\partial C_{\lambda\mu}}\right] X^{\varrho}_{,p} \nu_{\varrho} X^{\sigma}_{,q} \nu_{\sigma}, \\
&= D^{\alpha}_{\beta}{}^{\gamma\delta} \nu_{\gamma} \nu_{\delta},
\end{aligned} \tag{6.13}$$

where

$$D^{\alpha}_{\beta}{}^{\gamma\delta}(\boldsymbol{C}) = \delta^{\alpha}_{\beta} t^{\gamma\delta} + C_{\beta\lambda}\left(\frac{\partial t^{\alpha\gamma}}{\partial C_{\delta\lambda}} + \frac{\partial t^{\alpha\delta}}{\partial C_{\gamma\lambda}}\right). \tag{6.14}$$

Thus, for a given material, $D^{\alpha}_{\beta}{}^{\gamma\delta}$ is a given function of $\boldsymbol{C}$ and hence is independent of the local rotation. From (6.13) and (6.12) we see that the propagation condition (2.7) may be written in the form

$$(S^{\alpha}_{\beta} - \tilde{\varrho}\, U^2 \delta^{\alpha}_{\beta}) A^{\beta} = 0, \tag{6.15}$$

where we may[4] choose $\boldsymbol{A}$ in such a way that

$$A^{\beta} = X^{\beta}_{,k} a^k = \overset{-1}{R}{}^{\beta}_{k} \overset{\frac{1}{2}}{c}{}^k_m a^m, \tag{6.16}$$

$\boldsymbol{R}$ being the local rotation. The speeds of propagation are determined from (6.15) by the cubic equation

$$\det(S^{\alpha}_{\beta} - \tilde{\varrho}\, U^2 \delta^{\alpha}_{\beta}) = 0 \tag{6.17}$$

and hence are algebraic functions of the principal invariants of the tensor $\boldsymbol{S}(\boldsymbol{C}, \boldsymbol{\nu})$. The right proper vectors of $\boldsymbol{S}$ may be calculated from those of $\boldsymbol{Q}$ by means of (6.16); thus the acoustic axes may be obtained by transforming the right proper vectors of $\boldsymbol{S}$ like material elements in the deformation.

7. Isotropic materials

For an isotropic material the function $\mathbf{t}(\boldsymbol{C})$ is an isotropic function. By using the representation theorem for such functions it is easy to show that (6.1) may be reduced[1] to the form

$$t^k_m = f_0 \delta^k_m + f_1 B^k_m + f_2 B^k_p B^p_m, \tag{7.1}$$

where[2]

$$B^{km} \equiv g^{\alpha\beta} x^k_{,\alpha} x^m_{,\beta}, \tag{7.2}$$

and where the response coefficients f_0, f_1, f_2 are functions of the principal invariants I, II, III of $\boldsymbol{B}$. CAUCHY's first law of motion (2.4) may be written in the

[3] This analysis generalizes and simplifies that of TOUPIN & BERNSTEIN [1961, 1, §4]. It employs the identity

$$t^{qp} = \frac{\varrho}{\tilde{\varrho}} x^q_{,\gamma} x^p_{,\delta} t^{\gamma\delta}(\boldsymbol{C}), \tag{C}$$

which follows from (6.1) and Eq. (A) in footnote 3, p. 266. The result is equivalent to one stated and proved in a different way, again for the hyperelastic case, by TOLOTTI [1943, §1, p. 1]: The form of the quadric of polarization is independent of the local rotation.

[4] This possibility follows by decomposing the deformation into a rotation and a pure stretch; *cf.* CFT, §37.

[1] *E.g.*, §303 of CFT.

[2] $\boldsymbol{B}$ is the tensor denoted by $\mathbf{c}^{-1}$ in §29 of CFT.

more familiar form [CFT, Eq. (205.2)]

$$t^{km}{}_{,m} + \varrho f^k = \varrho \ddot{x}^k, \tag{7.3}$$

where ϱ is the density in the deformed material. By putting (7.1) into (7.3) we see that[3]

$$T^{km}{}_{pq} B^{pq}{}_{,m} + \varrho f^k = \varrho \ddot{x}^k, \tag{7.4}$$

where[4]

$$\begin{aligned} T^{km}{}_{pq} &\equiv \frac{\partial t^{km}}{\partial B^{pq}} = T^{mk}{}_{pq} = T^{km}{}_{qp}, \\ &= \tfrac{1}{2} f_1 (\delta^k_p \delta^m_q + \delta^m_p \delta^k_q) + \\ &+ \tfrac{1}{2} f_2 (\delta^k_p B^m_q + \delta^k_q B^m_p + \delta^m_p B^k_q + \delta^m_q B^k_p) + \\ &+ g^{km} \left[\frac{\partial f_0}{\partial I} g_{pq} + \frac{\partial f_0}{\partial II} (I g_{pq} - B_{pq}) + III \frac{\partial f_0}{\partial III} \overset{-1}{B}_{pq}\right] + \\ &+ B^{km} \left[\frac{\partial f_1}{\partial I} g_{pq} + \frac{\partial f_1}{\partial II} (I g_{pq} - B_{pq}) + III \frac{\partial f_1}{\partial III} \overset{-1}{B}_{pq}\right] + \\ &+ B^k_r B^{rm} \left[\frac{\partial f_2}{\partial I} g_{pq} + \frac{\partial f_2}{\partial II} (I g_{pq} - B_{pq}) + III \frac{\partial f_2}{\partial III} \overset{-1}{B}_{pq}\right], \end{aligned} \tag{7.5}$$

and, by (7.2),

$$B^{pq}{}_{,m} = X^\alpha{}_{,m} g^{\beta\gamma} (x^p{}_{,\gamma} x^q{}_{;\alpha\beta} + x^q{}_{,\gamma} x^p{}_{;\alpha\beta}). \tag{7.6}$$

If we calculate the jump of (7.4) at a singular surface of second order, by using (7.6) and (2.1) we again obtain the propagation condition (2.7), where $\boldsymbol{Q}$ now has the following explicit form:

$$Q^k{}_m = 2 \frac{\tilde{\varrho}}{\varrho} T^{kp}{}_{mq} B^{qs} n_p n_s. \tag{7.7}$$

By (7.5), $T^{kp}{}_{mq}$ is a function of $\boldsymbol{B}$ only; hence, for given $\boldsymbol{n}$, the acoustical tensor $\boldsymbol{Q}$ is a function of $\boldsymbol{B}$ only. Therefore, as is only natural, *in an isotropic medium not only the speeds of propagation but also the acoustic axes for waves travelling in a given direction are determined by the strain alone.*

A wave travelling down a principal axis of stress and strain in an isotropic medium may be called a *principal wave.* The results to be expected in this special case have been described in §1. To render explicit the statements made there, let $\boldsymbol{n}_1$ be a proper vector of $\boldsymbol{B}$, so that $B^r_s n^s_1 = v^2_1 n^r_1$, $\overset{-1}{B}{}^r_s n^s_1 = n^r_1/v^2_1$, where v_1 is the stretch in the direction of the principal axis corresponding to $\boldsymbol{n}_1$. Substituting (7.5) into (7.7) yields at once the following explicit form for the acoustical tensor[5]:

$$\begin{aligned} \frac{\varrho}{\tilde{\varrho}} v_1^{-2} Q^k{}_m(\boldsymbol{n}_1) &= (f_1 + v^2_1 f_2)\, \delta^k_m + f_2 B^k_m + \\ &+ 2 n^k_1 n^1_m \left\{\frac{1}{2} f_1 + v^2_1 f_2 + \sum_{\Gamma=0}^{2} v_1^{2\Gamma} \left[\frac{\partial f_\Gamma}{\partial I} + (v^2_2 + v^2_3) \frac{\partial f_\Gamma}{\partial II} + v^2_2 v^2_3 \frac{\partial f_\Gamma}{\partial III}\right]\right\}. \end{aligned} \tag{7.8}$$

---

[3] The approach here is that introduced by ERICKSEN [1953, 1] for more special circumstances.

[4] To effect the calculation, use Eqs. App. (38.16) of CFT.

[5] In effect, this tensor may be calculated (in the hyperelastic case) by combining Eqs. (4.8), (4.7), and (3.10) of the paper by HAYES & RIVLIN [1961, 2], but, perhaps hindered by an exuberance of formalism, they did not state the simple and conclusive results derived here. Earlier TOLOTTI [1943, Eqs. (18′) (19′)] had virtually calculated $\boldsymbol{Q}(\boldsymbol{n})$ for isotropic hyperelastic materials but, again apparently because of obscurity of notation and method, did not conclude anything except the independence of the acoustic axes from the local rotation.

Since the remaining principal axes of $\boldsymbol{B}$ are normal to $\boldsymbol{n}_1$, we see by inspection of (7.8) that they are also principal axes of $\boldsymbol{Q}(\boldsymbol{n}_1)$; that is, *the acoustic axes for principal waves coincide with the principal axes.* In particular, *every principal wave is either longitudinal or transverse.*

By using (2.10) we may now calculate the speeds of propagation exactly and explicitly[6]. Setting $\boldsymbol{a}=\boldsymbol{n}_1$, we obtain the squared speed U_{11}^2 of longitudinal waves along the principal axis in which the stretch is v_1:

$$\frac{\varrho\, U_{11}^2}{2v_1^2} = f_1 + 2v_1^2 f_2 + \sum_{\Gamma=0}^{2} v_1^{2\Gamma}\left[\frac{\partial f_\Gamma}{\partial I} + (v_2^2+v_3^2)\frac{\partial f_\Gamma}{\partial II} + v_2^2 v_3^2 \frac{\partial f_\Gamma}{\partial III}\right], \tag{7.9}$$

while for transverse waves with amplitude parallel to $\boldsymbol{n}_2$ the squared speed U_{12}^2 is given by

$$\frac{\varrho\, U_{12}^2}{v_1^2} = f_1 + (v_1^2+v_2^2)\, f_2 . \tag{7.10}$$

The formulae (7.9) and (7.10) give the ***explicit and exact speeds of propagation of principal waves.*** ERICKSEN[7] has pointed out that they assume a simpler form if expressed in terms of the relations giving the principal stresses t_a as functions of the principal stretches:

$$t_1 = f_0 + f_1 v_1^2 + f_2 v_1^4, \tag{7.11}$$

as follows from (7.1). Since

$$\frac{\partial f_\Gamma}{\partial v_1^2} = \frac{\partial f_\Gamma}{\partial I} + (v_2^2+v_3^2)\frac{\partial f_\Gamma}{\partial II} + v_2^2 v_3^2\frac{\partial f_\Gamma}{\partial III}, \tag{7.12}$$

from (7.11) we see that the right-hand side of (7.9) is equal to $\partial t_1/\partial v_1^2$. Hence[8]

$$\varrho\, U_{11}^2 = \frac{\partial t_1}{\partial \log v_1}, \tag{7.13}$$

while from (7.10) it follows at once that

$$\frac{\varrho\, U_{12}^2}{v_1^2} = \frac{t_1 - t_2}{v_1^2 - v_2^2}. \tag{7.14}$$

Since these elegant formulae do not make direct use of the response coefficients f_Γ, they give insight into the general nature of wave propagation in all kinds of elastic materials.

---

[6] These results generalize a long sequence of special formulae for various kinds of materials in various circumstances. Some of the relevant literature, which is mainly physical, is cited by TOUPIN & BERNSTEIN [1961, 1].

The formula (7.10) is similar to one given by ERICKSEN [1953, 1, Eq. (4.6)] for incompressible hyperelastic materials. The results above, however, do not seem to be obvious from his, because the condition of incompressibility by itself renders all waves transversal, and thus for incompressible materials the whole picture of wave propagation is not really representative of the general case.

[7] Letter dated July 13, 1961. {In fact (7.13) and (7.14) are closely related to results that TRUESDELL [1956, § 8] attributed to unpublished work of ERICKSEN & TOUPIN.}

[8] Here t_1 is regarded as a function of the three *independent* variables $\log v_1$, $\log v_2$ and $\log v_3$. For special families of deformations a result of different form will follow. For example, if only deformations such that $v_1=v_2=v_3$ are considered, we obtain a result of the form (8.4) below.

By using (7.10) we see that

$$\frac{\varrho}{v_1^2}(U_{12}^2 - U_{13}^2) = (v_2^2 - v_3^2) f_2. \tag{7.15}$$

From this formula follows the ***fundamental theorem on transverse principal waves:*** *In an isotropic elastic material such that $f_2 = 0$ for the strain in question, both kinds of transverse waves that may travel down a given principal axis of strain have the same absolute speed of propagation; if, however, $f_2 \neq 0$, then these two kinds of transverse waves travel at the same absolute speed if and only if the corresponding principal stretches are equal; and in this latter case, any transverse amplitude is possible, and all these transverse waves have the same absolute speeds of propagation.* From (7.15), moreover, we may interpret the sign of the function f_2. Excluding the case when $v_2 = v_3$, we see that *if $f_2 < 0$, transverse waves with amplitudes parallel to the axis of lesser transverse stretch travel at greater absolute speed than the others; if $f_2 > 0$, at lesser*[9].

From (7.9) it is clear that $U_{11}^2 = U_{22}^2$ if $v_1 = v_2$. A converse condition is immediate from (7.13), leading to the ***fundamental theorem on longitudinal principal waves:*** *In an isotropic elastic material, the squared speeds of longitudinal waves in the directions of two equal principal stretches are equal; conversely, if $U_{11}^2 = U_{22}^2$, it is necessary that*[10]

$$\frac{\partial t_1}{\partial \log v_1} = \frac{\partial t_2}{\partial \log v_2}. \tag{7.16}$$

These results show that the state of affairs in the linearized theory, where all transverse waves travel at one absolute speed and all longitudinal waves at another, is not at all typical for elastic materials.

8. Isotropic materials subject to hydrostatic pressure

Only the case of hydrostatic stress partakes of the simplicity to which the linearized theory has accustomed us. Indeed, if $v_1 = v_2 = v_3 = v$, then every direction is a principal axis of stress and strain, so that all waves are principal waves, while (7.9) and (7.10) show that $U_{11}^2 = U_{22}^2 = U_{33}^2 = U_{\parallel}^2$, say, and $U_{12}^2 = U_{21}^2 = \cdots = U_{32}^2 = U_{\perp}^2$, say. That is, *in a state of hydrostatic pressure, longitudinal and transverse waves may propagate in any direction*[1]; *at each point, all longitudinal waves have the same absolute speed, and all transverse waves have the same absolute speed.* To calculate these speeds, we note that since $I = 3v^2$, $II = 3v^4$, $III = v^6$, it follows that

$$\frac{\partial}{\partial I} + 2v^2 \frac{\partial}{\partial II} + v^4 \frac{\partial}{\partial III} = \frac{1}{3}\frac{d}{dv^2} \tag{8.1}$$

for the class of deformations being considered. Moreover, by (7.1), $t^k_m = -p\,\delta^k_m$ where $p = p(\varrho)$. From (7.9) and (7.10) it is now easy to show that

$$U_{\parallel}^2 = \frac{4}{3}\frac{v^5}{\varrho}(f_1 + 2v^2 f_2) + \frac{dp}{d\varrho}, \tag{8.2}$$

$$U_{\perp}^2 = \frac{v^5}{\varrho}(f_1 + 2v^2 f_2). \tag{8.3}$$

---

[9] For rubbers $f_2 < 0$ according to the results of experiments. The sign of the more frequently used coefficient h_{-1}, which is introduced in § 11, is the same as that of f_2.

[10] By calculating $U_{11}^2 - U_{22}^2$ from (7.9) it is a trivial matter to express this condition in terms of the response coefficients f_Γ, but the resulting formula is too elaborate to be useful.

[1] Recall the caution expressed in § 5.

By eliminating $f_1 + 2v^2 f_2$ between these two results, we derive the following not inelegant *universal relation* between the two wave speeds and the bulk modulus in any state of hydrostatic pressure:

$$U_{\|}^2 = \frac{4}{3} U_{\perp}^2 + \frac{dp}{d\varrho}. \tag{8.4}$$

Two special cases are already familiar. First, for an ideal fluid we have $f_1 = f_2 = 0$, while $f_0 = g(III) = -p(\varrho)$, say. By (8.3), $U_{\perp} = 0$ in this case[2], while (8.4) reduces to the classical formula for the speed of sound in fluids. Second, in the linearized theory of elasticity, (8.4) reduces to a trivial identity connecting the two squared wave-speeds, *viz* $(\lambda + 2\mu)/\varrho$, μ/ϱ, with the bulk modulus, $(\lambda + \frac{2}{3}\mu)/\varrho$.

It is reasonable to assume that $dp/d\varrho > 0$ in all circumstances. Thus from (8.4) we see that

$$U_{\|}^2 > \tfrac{4}{3} U_{\perp}^2 \tag{8.5}$$

for any state of hydrostatic pressure in any elastic material. In particular, *longitudinal waves always propagate faster than transverse waves*, provided the latter can exist at all. Whether or not $U_{\perp}$ is real is a more delicate matter. Some conditions of reality are discussed in §10. For the time being, however, we content ourselves with interpreting (8.4) as follows: *In a state of hydrostatic stress, if transverse waves may propagate at all, so may longitudinal waves, and then*

$$U_{\|}^2 > \frac{dp}{d\varrho}; \tag{8.6}$$

that is, if transverse waves may propagate, the squared speed of longitudinal waves is always greater than the squared wave-speed according to the hydrodynamics of an ideal fluid with the same law of compression. If, however, $U_{\perp}^2 < 0$, then

$$U_{\|}^2 < \frac{dp}{d\varrho}. \tag{8.7}$$

Therefore, if neither material transverse singularities nor transverse waves are possible, the squared speed of longitudinal waves is either less than that of sound waves in a fluid with the same law of compression, or is zero, or is imaginary. The case when $U_{\|}^2 < 0$ cannot be excluded *a priori*.

The foregoing results on the speeds of waves in elastic materials subject to great hydrostatic pressure may be not devoid of interest to seismologists.

9. Determination of the stress-strain relations from wave speeds

When the form of the stress-strain relations is known, formulae such as (7.9) and (7.10) give unique values for the squared speeds of the 9 possible kinds of principal waves. Conversely, if these 9 squared speeds are known as functions of the three principal stretches v_a, as for example from the results of experiments, then the three response coefficients f_0, f_1, and f_2 are uniquely determined.

Indeed, they are overdetermined. From (7.9) and (7.10) we may derive a sequence of *universal relations* connecting the wave speeds, relations which must hold in any isotropic elastic material regardless of the forms of f_0, f_1, and f_2. Only if these relations are satisfied can there exist functions f_0, f_1, and f_2 such

[2] Thus transverse singular surfaces of second or higher order cannot propagate in an ideal fluid. Other remarks on this case are given in footnote 4, p. 281, below.

that (7.9) and (7.10) hold. Thus these universal relations serve as *conditions of compatibility* for measured values of the wave speeds: Unless these relations be satisfied by the data, it is not consistent with the theory of isotropic elasticity[1].

Compatibility of transverse waves. From (7.10) we derive at once 2 sets of 3 universal relations:

$$\frac{U_{12}^2}{v_1^2} = \frac{U_{21}^2}{v_2^2}, \qquad \frac{U_{23}^2}{v_2^2} = \frac{U_{32}^2}{v_3^2}, \qquad \frac{U_{31}^2}{v_3^2} = \frac{U_{13}^2}{v_1^2}, \tag{9.1}$$

$$\begin{aligned} \frac{1}{v_1^2-v_3^2}\left(\frac{U_{12}^2}{v_1^2} - \frac{U_{23}^2}{v_2^2}\right) = \frac{1}{v_2^2-v_1^2}\left(\frac{U_{23}^2}{v_2^2} - \frac{U_{31}^2}{v_3^2}\right) = \frac{1}{v_3^2-v_2^2}\left(\frac{U_{31}^2}{v_3^2} - \frac{U_{12}^2}{v_1^2}\right), \\ \equiv \varphi(v_1, v_2, v_3), \end{aligned} \tag{9.2}$$

where φ is a symmetric function of its arguments. Of these 6 conditions, the first 3 are independent, but from any 1 of the last 3 the other 2 may be derived by use of the first set, so in all there are 4 independent conditions. If these are satisfied, the functions f_1 and f_2 are uniquely determined by relations of the type (7.10). For example,

$$\begin{aligned} f_2 &= \varrho\,\varphi = \varrho\,\frac{U_{12}^2 - U_{13}^2}{v_1^2(v_2^2-v_3^2)}, \\ f_1 &= \frac{1}{3}\varrho\left[\frac{U_{12}^2}{v_1^2} + \frac{U_{23}^2}{v_2^2} + \frac{U_{31}^2}{v_3^2} - 2I\varphi\right], \\ &= \varrho\,\frac{(v_2^2 - I)\,U_{12}^2 - (v_3^2 - I)\,U_{13}^2}{v_1^2(v_2^2-v_3^2)}. \end{aligned} \tag{9.3}$$

In each case the second form follows from the first by use of the identities (9.1) and (9.2). From the form of the results we see that *in an isotropic elastic material, the response coefficients f_1 and f_2 are uniquely determined by the speeds of the two kinds of transverse principal waves corresponding to any one principal axis.* The experimenter willing to take compatibility for granted may content himself with measuring U_{12}^2 and U_{13}^2 as functions of v_1, v_2, v_3.

Compatibility of longitudinal waves. Let us assume that the conditions of compatibility for the transverse principal waves are satisfied, so that f_1 and f_2 are uniquely determined. We now consider the compatibility of the speeds of the 3 longitudinal waves, from which the single function f_0 is to be determined.

If we set

$$A_1 \equiv \frac{1}{2}\varrho\,\frac{U_{11}^2}{v_1^2} - f_1 - 2v_1^2 f_2 - \sum_{\Gamma=1}^{2} v_1^{2\Gamma}\left[\frac{\partial f_\Gamma}{\partial I} + (v_2^2+v_3^2)\frac{\partial f_\Gamma}{\partial II} + v_2^2 v_3^2 \frac{\partial f_\Gamma}{\partial III}\right], \tag{9.4}$$

etc., then because of (9.3) we can regard the 3 quantities A_a as explicitly given functions of $U_{12}^2, U_{13}^2, U_{11}^2, U_{22}^2, U_{33}^2, v_1, v_2, v_3$, and we are to find f_0 from 3 relations

---

[1] Toupin & Bernstein [1961, 1, §4] were the first to consider problems of this kind. They derived necessary and sufficient conditions for the linearized theory of hyperelastic materials. Earlier Hughes & Kelly [1953, 2, p. 1146], after obtaining more than sufficient formulae for determining the second-order elasticities of isotropic hyperelastic materials, had remarked that "any additional measurements give checks on the theory".

The results I present above for the finite theory are sufficient only as regards the nine kinds of principal waves. If waves travelling in other directions are to be considered, additional conditions of compatibility must be satisfied.

of the type

$$A_1 = \frac{\partial f_0}{\partial I} + (v_2^2 + v_3^2)\frac{\partial f_0}{\partial II} + v_2^2 v_3^2 \frac{\partial f_0}{\partial III}. \tag{9.5}$$

If the three principal stretches v_a are distinct, we may solve (9.5) as a linear system, obtaining the following results:

$$\begin{aligned}
\frac{\partial f_0}{\partial I} &= \frac{A_1 v_1^4 (v_2^2 - v_3^2) + A_2 v_2^4 (v_3^2 - v_1^2) + A_3 v_3^4 (v_1^2 - v_2^2)}{(v_1^2 - v_2^2)(v_2^2 - v_3^2)(v_3^2 - v_1^2)} \equiv B_1,\\
\frac{\partial f_0}{\partial II} &= -\frac{A_1 v_1^2 (v_2^2 - v_3^2) + A_2 v_2^2 (v_3^2 - v_1^2) + A_3 v_3^2 (v_1^2 - v_2^2)}{(v_1^2 - v_2^2)(v_2^2 - v_3^2)(v_3^2 - v_1^2)} \equiv B_2,\\
\frac{\partial f_0}{\partial III} &= \frac{A_1 (v_2^2 - v_3^2) + A_2 (v_3^2 - v_1^2) + A_3 (v_1^2 - v_2^2)}{(v_1^2 - v_2^2)(v_2^2 - v_3^2)(v_3^2 - v_1^2)} \equiv B_3.
\end{aligned} \tag{9.6}$$

The quantities B_a are known functions of v_1, v_2, v_3; they must be symmetric functions and hence equal to functions of I, II, III. In order that f_0 may be determined from the system (9.6), the functions B_a must satisfy the following conditions of integrability:

$$\frac{\partial B_1}{\partial II} = \frac{\partial B_2}{\partial I}, \qquad \frac{\partial B_2}{\partial III} = \frac{\partial B_3}{\partial II}, \qquad \frac{\partial B_1}{\partial III} = \frac{\partial B_3}{\partial I}. \tag{9.7}$$

If these conditions are satisfied, then

$$f_0 = \int (B_1\, dI + B_2\, dII + B_3\, dIII). \tag{9.8}$$

A summary of these results is given by the ***compatibility theorem:*** *Let the* 9 *squared speeds of propagation of principal waves be known functions of the principal stretches. This data is consistent with the theory of isotropic elastic materials if and only if:*

(a) *The quantities* φ, B_1, B_2, B_3, *as defined by* $(9.2)_3$ *and* (9.6), *are symmetric functions of the principal stretches.*

(b) *The* 4 *algebraic conditions* (9.1) *and* $(9.2)_1$ *and the* 3 *differential conditions* (9.7) *are satisfied.*

If conditions (a) *and* (b) *hold, the data determines a unique form for the stress-strain relations by means of* (9.3) *and* (9.8).

10. Conditions that the waves be real

As yet, we have not taken up the question of whether the speeds and corresponding amplitudes are real. Only real waves have physical significance. The results so far stated are carefully phrased so as to refer, for the most part, only to necessary conditions. In the generality maintained thus far, *it is possible that no waves can exist.* For $\boldsymbol{Q}(\boldsymbol{n})$, being an arbitrary 3×3 matrix, may have but one real right proper vector, and the corresponding proper number may be negative, leading to a purely imaginary value of U.

Various inequalities have been proposed so as to insure that the response function $\mathfrak{h}_k^{\alpha}$ in (2.3) describe a material not altogether unreasonable for physical applications[1]. All of these turn out to have some bearing on the propagation

[1] A review of this question, which I proposed in a lecture [1956] before the Gesellschaft für angewandte Mathematik und Mechanik in Berlin in 1955, is given by TOUPIN [1961, 3].

of waves, and in this context we are now in a position to read off results more general than any heretofore known.

First, for the hyperelastic case HADAMARD [1903, ¶ 269] gave a definition of stability from which he proved that in a stable state all speeds of propagation are real. I shall not discuss this condition here[1] because it is not clear how to extend HADAMARD's definition of stability to general elastic materials, when there need be no strain energy.

In the hyperelastic case, however, HADAMARD's criterion implies that the equations of equilibrium are *strongly elliptic*, and a condition to this end we can easily formulate and discuss in full generality from (2.6); namely, the *S-E condition* is

$$A_{(km)}{}^{(\alpha\beta)}\, l^k\, l^m\, \lambda_\alpha\, \lambda_\beta > 0 \tag{10.1}$$

for arbitrary vectors $\boldsymbol{l}$ and $\boldsymbol{\lambda}$. From (2.8) we see that an equivalent condition is

$$Q_{(km)}(\boldsymbol{n})\, l^k\, l^m > 0 \tag{10.2}$$

for arbitrary $\boldsymbol{n}$ and $\boldsymbol{l}$. That is, *the S-E condition is necessary and sufficient that the symmetric part of the acoustical tensor be positive-definite, for every direction of propagation.* From (2.10) it thus follows that $U^2 > 0$ if $\boldsymbol{a}$ is real. That is, *if the S-E condition holds, the squared speed of propagation corresponding to every real acoustic axis is positive.* In particular, since for given $\boldsymbol{n}$ there is at least one real amplitude $\boldsymbol{a}$, *the S-E condition implies that for any direction of propagation there exists at least one real amplitude with positive speed.*

Since, however, the *S-E* condition restricts only the symmetric part of $\boldsymbol{Q}(\boldsymbol{n})$, it cannot guarantee that in general there are three real acoustic axes. In the important special case of principal waves in an isotropic material, we have exhibited three real axes. By the results in §6, then, *the S-E condition implies that the squared speeds of all principal waves in an isotropic material are positive.*

The foregoing results, except for the last one, which refers to isotropic materials, are independent of the principle of material indifference. When this principle is satisfied, from (6.8) we see that (10.1) may be expressed in the following equivalent form:

$$B^{kpmq}\, l_k\, l_m\, L_p\, L_q > 0 \tag{10.3}$$

for arbitrary vectors $\boldsymbol{l}$ and $\boldsymbol{L}$. A formally similar but in fact different condition is

$$B^{kpmq}\, S_{kp}\, S_{mq} > 0 \tag{10.4}$$

---

[1] It must be added, however, that the criticism by SIGNORINI and TOLOTTI [1943] entirely misses the point of HADAMARD's work by omitting any mention of the question of stability. As shown in the text above, it is entirely obvious that for fully general stress-strain relations there need be no possibility of real waves. Moreover, the very special stress-strain relations derived by TOLOTTI [1943, Eq. (3)] do not exhaust the possible isotropic hyperelastic materials in which all speeds of propagation are real, for TOLOTTI imposes at the outset the unduly restrictive additional condition that one acoustical axis be longitudinal for any direction of propagation, while transverse waves may have *any* perpendicular amplitude. As to be expected, he succeeds in showing that scarcely any materials are of this kind.

HADAMARD's analysis of stability was questioned by DUHEM [1905, Ch. III, § IV]. A new proof of HADAMARD's stability theorem was given by C. CATTANEO, "Su un teorema fondamentale nella teoria delle onde di discontinuità", Atti Accad. Lincei Rend. Cl. Sci. Fis. Mat. Nat. (8) 1, 66—72, 728—734 (1946).

for all symmetric non-vanishing tensors $\boldsymbol{S}$. For hyperelastic materials, a condition of this type is equivalent to one proposed by COLEMAN & NOLL[3] on the basis of a compelling thermodynamic argument, supported by important corollaries in statics. We shall call (10.4) *the generalized C-N condition.* Unlike the *S-E* condition, it cannot be expressed entirely in terms of the acoustical tensor.

It is difficult to draw general conclusions from (10.4). From examples[4] it is known that neither the generalized *C-N* condition nor the *S-E* condition implies the other for all materials and all strains. In the special case when we put $S_{kp}=n_k n_p$, by using (6.11) we derive at once the necessary condition

$$Q^{km}(\boldsymbol{n})\, n_k\, n_m > 0. \tag{10.5}$$

In general, $\boldsymbol{n}$ is not a right proper vector of $\boldsymbol{Q}(\boldsymbol{n})$; if it is, from (10.5) and (2.10) we see that the corresponding speed of propagation is real. That is, *the generalized C-N condition insures that if a longitudinal amplitude is possible, the corresponding squared speed is positive.*

In the isotropic case, ERICKSEN's formulae (7.13) and (7.14) make possible the statement of simple and immediate criteria, which may be summarized in the ***fundamental theorem on the speeds of principal waves:*** *In an isotropic material the squared speeds of the principal longitudinal waves are positive if and only if each principal tension is an increasing function of the corresponding principal stretch when the other principal stretches are held constant; the squared speeds of the principal transverse waves are positive if and only if the greater principal tension occurs always in the direction of the greater principal stretch.*

As has already been shown, the *S-E* condition implies that all squared wave speeds are positive; hence it implies the two statical conditions just stated. The first of these is implied also by the generalized *C-N* condition; to verify this fact,

---

[3] For the connection with the work of COLEMAN & NOLL [1959, 1], the reader is referred to the paper of TOUPIN & BERNSTEIN [1961, 1, §3]. As explained there, the requirements of COLEMAN & NOLL allow equality to replace inequality in (10.4) for certain exceptional strains, but in the interest of simpler statements of theorems we do not consider these.

[4] TOUPIN & BERNSTEIN [1960, 1, §3] construct a simple example of an isotropic hyperelastic solid which when subject to hydrostatic pressures within a certain range satisfies the *C-N* condition but does not satisfy the *S-E* condition.

As they remark, an ideal fluid for which $p>0$ and $dp/d\varrho>0$ serves as an example of a material in which the *C-N* condition is satisfied for all strains but the *S-E* condition for none. To verify this fact from the present point of view it suffices to note that $f_0=g(III)=-p(\varrho)$, $f_1=f_2=0$, so that (7.7) yields

$$Q^k_m(\boldsymbol{n})=2\frac{\tilde{\varrho}}{\varrho}\,III\,\frac{\partial g}{\partial III}\,n^k\,n_m=\tilde{\varrho}\,\frac{dp}{d\varrho}\,n^k\,n_m. \tag{D}$$

As in all isotropic materials subject to hydrostatic pressure, any transverse amplitude is possible, *but here the corresponding speed of propagation is zero.* That is, in the notation of §8, $U_\perp=0$. From (2.1) and (3.1) it follows that *in an ideal fluid, the only possible transverse singular surfaces of order greater than* 1 *are material surfaces across which only certain purely spatial derivatives such as* $x^k{}_{,\alpha\beta}$, $x^k{}_{,\alpha\beta\gamma}$, etc. *suffer discontinuities* [*cf.* CFT, §190]. The *C-N* condition, as was shown in the researches of COLEMAN & NOLL, is sufficient that $dp/d\varrho>0$ and hence that $U_\parallel^2>0$, by (8.4). Since $U_\perp=0$, the *S-E* condition can never be satisfied. Conversely, the condition that the squared speeds of longitudinal waves be positive, since it is equivalent to $dp/d\varrho>0$, is weaker than the *C-N* condition, which implies that $p>0$ also.

we need only notice that, as was shown in §7, a longitudinal amplitude is always possible for a principal wave in an isotropic material, so that the italicized statement following (10.5) becomes applicable. Thus *the generalized C-N condition implies that the squared speeds of all principal longitudinal waves in an isotropic material are positive*[5].

The generalized *C-N* condition is not so intimately connected with transverse waves. The second statical condition in the last theorem, namely, that the greater principal tension occurs in the direction of the greater principal stretch, was proposed and explored by BAKER & ERICKSEN [1954] and may be called the *B-E condition*[6]. Merely rephrasing a result already proved, we may assert that *the B-E condition is necessary and sufficient that the squared speeds of all principal transverse waves be positive*[7]. It is easy to show directly that for isotropic hyperelastic[8] materials the *C-N* condition implies the *B-E* condition if all principal tensions are positive. Hence *the C-N condition implies that in a state of pure tension in an isotropic hyperelastic material, the squared speeds of all principal waves are positive.*

11. Waves according to the second-order theory of elasticity for isotropic materials*

We now apply the foregoing results to the determination of second-order elastic moduli from the speeds of waves. To this end we write (7.1) in the form[1]

$$t^k_m = h_0 \delta^k_m + h_1 B^k_m + h_{-1} \overset{-1}{B}{}^k_m, \qquad (11.1)$$

where the coefficients $h_\Gamma(I, II, III)$ are related as follows to the coefficients f_Γ used previously:

$$f_0 = h_0 + \frac{II}{III} h_{-1}, \quad f_1 = h_1 - \frac{I}{III} h_{-1}, \quad f_2 = \frac{I}{III} h_{-1}. \qquad (11.2)$$

The classical tensor $\boldsymbol{E}$ for measuring finite strain is given by

$$E^\alpha_\beta \equiv \tfrac{1}{2}(C^\alpha_\beta - \delta^\alpha_\beta). \qquad (11.3)$$

---

* A treatment of this subject from a slightly different viewpoint will appear in the proceedings of the Symposium on Second-Order Effects in Elasticity, Plasticity and Fluid Dynamics, to be held at Haifa next Easter.

[5] This theorem is due to TOUPIN [1961, 3, Th. XII]. While his argument goes along correct lines, the undetermined form for $\boldsymbol{Q}(\boldsymbol{n})$ he in effect assumes as starting point is not sufficiently general, as may be seen from the explicit formula (7.8), above.

[6] In fact BAKER & ERICKSEN allow inequality to be replaced by equality if $v_a = v_b$, but for ease of stating the result we leave this possibility out of account.

[7] Hence follows as a trivial corollary a result apparently first published, and only for hyperelastic materials, by HAYES & RIVLIN [1961, 2]: *The S-E condition implies the B-E condition.* This fact is easily proved directly from the definitions, without reference to the theory of waves.

The theorem in the text above is close to one proved by ERICKSEN [1953, 1, § 4] for incompressible hyperelastic materials. {Indeed, it is extremely close to a theorem attributed to ERICKSEN & TOUPIN by TRUESDELL [1956, § 8].}

[8] I believe this result holds for all isotropic elastic materials but have not yet succeeded in constructing a proof.

[1] For the purposes of the present paper it would be easier to remain with the coefficients f_Γ, but the coefficients h_Γ are those employed in most of the modern researches on other problems in elasticity.

Hence the principal invariants of $\boldsymbol{E}$ are related to those of $\boldsymbol{B}$ by the following identities [CFT, Eqs. (31.5)]:

$$\begin{aligned} I_{\boldsymbol{E}} &= \tfrac{1}{2}(I-3), & I &= 3+2I_{\boldsymbol{E}}, \\ II_{\boldsymbol{E}} &= \tfrac{1}{4}(II-2I+3), & II &= 3+4I_{\boldsymbol{E}}+4II_{\boldsymbol{E}}, \\ III_{\boldsymbol{E}} &= \tfrac{1}{8}(III-II+I-1), & III &= 1+2I_{\boldsymbol{E}}+4II_{\boldsymbol{E}}+8III_{\boldsymbol{E}}. \end{aligned} \tag{11.4}$$

Let the response coefficients h_Γ be expanded in power series in the invariants of $\boldsymbol{E}$:

$$h_\Gamma/\mu = \alpha_{\Gamma 0} + \alpha_{\Gamma 1} I_{\boldsymbol{E}} + \alpha_{\Gamma 2} I_{\boldsymbol{E}}^2 + \alpha_{\Gamma 3} II_{\boldsymbol{E}} + \cdots. \tag{11.5}$$

The extensions δ_a are defined in terms of the stretches v_a as follows:

$$v_a = 1 + \delta_a. \tag{11.6}$$

Our objective is to express the wave speeds approximately as linear functions of the δ_a, with coeffcients determined by the $\alpha_{\Gamma\Delta}$ occurring in the assumed expansions (11.5). It turns out that in the results the $\alpha_{\Gamma\Delta}$ occur only in the following combinations[2]:

$$\begin{aligned} \frac{\lambda}{\mu} \equiv \alpha_1 &\equiv \alpha_{01} + \alpha_{11} + \alpha_{-11}, \\ 1 &\equiv \alpha_{10} - \alpha_{-10}, \\ \alpha_3 &\equiv \alpha_{02} + \alpha_{12} + \alpha_{-12}, \\ \alpha_4 &\equiv \alpha_{03} + \alpha_{13} + \alpha_{-13}, \\ \alpha_5 &\equiv 2(\alpha_{11} - \alpha_{-11}), \\ \alpha_6 &\equiv 4\alpha_{-10}, \end{aligned} \tag{11.7}$$

where λ and μ are the classical linear moduli.

Let us introduce the notation $\doteq$ to indicate equality to within an error which is $O(\delta_1^2+\delta_2^2+\delta_3^2)$ as $\delta_1^2+\delta_2^2+\delta_3^2 \to 0$. Let ϑ be the linear approximation to the cubical dilation:

$$\vartheta \equiv \delta_1 + \delta_2 + \delta_3. \tag{11.8}$$

Then

$$I_{\tilde{\boldsymbol{E}}} \doteq \vartheta. \tag{11.9}$$

From (7.10), by use of (11.5) and (11.7), we see that

$$\begin{aligned} \frac{\varrho U_{12}^2}{\mu} &\doteq \frac{1}{\mu}(1+2\delta_1)\,h_1 - \frac{1}{\mu}(1-2\delta_2)\,h_{-1}, \\ &\doteq (1+2\delta_1)(\alpha_{10}+\alpha_{11}\vartheta) - (1-2\delta_1)(\alpha_{-10}+\alpha_{-11}\vartheta), \\ &= (1+2\delta_1)(\alpha_{10}-\alpha_{-10}) + (\alpha_{11}-\alpha_{-11})\,\vartheta + 2\alpha_{-10}(\delta_1+\delta_2), \\ &= 1 + 2\delta_1 + \tfrac{1}{2}\alpha_5\vartheta + \tfrac{1}{2}\alpha_6(\delta_1+\delta_2). \end{aligned} \tag{11.10}$$

Since

$$\varrho_0 \doteq \varrho(1+\vartheta), \tag{11.11}$$

---

[2] To the same order of approximation, the stress-strain relations are of the form

$$t^k_m/\mu = \alpha_1 I_{\boldsymbol{E}}\,\delta^k_m + 2E^k_m + (\alpha_3 I^2_{\tilde{\boldsymbol{E}}} + \alpha_4 II_{\tilde{\boldsymbol{E}}})\,\delta^k_m + \alpha_5 I_{\tilde{\boldsymbol{E}}}\,\tilde{E}^k_m + \alpha_6 \tilde{E}^k_p \tilde{E}^p_k, \tag{E}$$

where $\tilde{\boldsymbol{E}}$ is the classical linear tensor of infinitesimal strain, but this fact is not used in the analysis.

we may express (11.10) in the alternative form

$$\frac{\varrho_0 U_{12}^2}{\mu} \doteq 1 + 2\delta_1 + \left(1 + \frac{1}{2}\alpha_5\right)\vartheta + \frac{1}{2}\alpha_6(\delta_1 + \delta_2), \tag{11.12}$$

giving the *second-order speeds of transverse principal waves.*

It is not so easy to get the corresponding results for longitudinal waves. First we write (7.9) in the form

$$\frac{\varrho U_{11}^2}{2v_1^2} = A + B, \tag{11.13}$$

where

$$\begin{aligned} A &\equiv f_1 + 2v_1^2 f_2, \\ B &\equiv \sum_{\Gamma=0}^{2} v_1^{2\Gamma}\left[\frac{\partial f_\Gamma}{\partial I} + (v_2^2 + v_3^2)\frac{\partial f_\Gamma}{\partial II} + v_2^2 v_3^2 \frac{\partial f_\Gamma}{\partial III}\right]. \end{aligned} \tag{11.14}$$

Then from (11.2), (11.5) and (11.7) it follows that

$$\begin{aligned} \frac{A}{\mu} &= \frac{1}{\mu}\left[h_1 + \frac{v_1^2 - v_2^2 - v_3^2}{v_1^2 v_2^2 v_3^2} h_{-1}\right], \\ &\doteq \alpha_{10} + \alpha_{11}\vartheta + \frac{-1-2\vartheta+4\delta_1}{1+2\vartheta}(\alpha_{-10} + \alpha_{-11}\vartheta), \\ &\doteq 1 + \tfrac{1}{2}\alpha_5\vartheta + \alpha_6\delta_1. \end{aligned} \tag{11.15}$$

Also

$$\begin{aligned} B &\doteq \frac{\partial f_0}{\partial I} + 2(1+\vartheta-\delta_1)\frac{\partial f_0}{\partial II} + (1+2\vartheta-2\delta_1)\frac{\partial f_0}{\partial III} + \\ &\quad + (1+2\delta_1)\left[\frac{\partial f_1}{\partial I} + 2(1+\vartheta-\delta_1)\frac{\partial f_1}{\partial II} + (1+2\vartheta-2\delta_1)\frac{\partial f_1}{\partial III}\right] + \\ &\quad + (1+4\delta_1)\left[\frac{\partial f_2}{\partial I} + 2(1+\vartheta-\delta_1)\frac{\partial f_2}{\partial II} + (1+2\vartheta-2\delta_1)\frac{\partial f_2}{\partial III}\right], \\ &\doteq \left(\frac{\partial}{\partial I} + 2\frac{\partial}{\partial II} + \frac{\partial}{\partial II}\right)(f_0 + f_1 + f_2) + \\ &\quad + 2(\vartheta - \delta_1)\left(\frac{\partial}{\partial II} + \frac{\partial}{\partial III}\right)(f_0 + f_1 + f_2) + \\ &\quad + 2\delta_1\left(\frac{\partial}{\partial I} + 2\frac{\partial}{\partial II} + \frac{\partial}{\partial III}\right)(f_1 + 2f_2). \end{aligned} \tag{11.16}$$

From (11.4) we see that

$$\frac{\partial}{\partial I} + 2\frac{\partial}{\partial II} + \frac{\partial}{\partial III} = \frac{1}{2}\frac{\partial}{\partial I_E}, \qquad \frac{\partial}{\partial II} + \frac{\partial}{\partial III} = \frac{1}{4}\frac{\partial}{\partial II_E}. \tag{11.17}$$

Hence (11.16) becomes

$$B \doteq \frac{1}{2}\frac{\partial}{\partial I_E}(f_0 + f_1 + f_2) + \frac{1}{2}(\vartheta - \delta_1)\frac{\partial}{\partial II_E}(f_0 + f_1 + f_2) + \delta_1\frac{\partial}{\partial I_E}(f_1 + 2f_2). \tag{11.18}$$

Now by (11.2) and (11.4),

$$f_0 + f_1 + f_2 = h_0 + h_1 + h_{-1} - \frac{8III_E}{1+2I_E+4II_E+8III_E}h_{-1}. \tag{11.19}$$

Hence

$$\begin{aligned} \frac{\partial}{\partial I_E}(f_0 + f_1 + f_2) &\doteq \frac{\partial}{\partial I_E}(h_0 + h_1 + h_{-1}), \\ &\doteq \tfrac{1}{2}\mu(\alpha_{01} + \alpha_{11} + \alpha_{-11}) + \mu(\alpha_{02} + \alpha_{12} + \alpha_{-12})\vartheta, \\ &= \mu(\tfrac{1}{2}\alpha_1 + \alpha_3\vartheta). \end{aligned} \tag{11.20}$$

Likewise

$$(\vartheta-\delta_1)\frac{\partial}{\partial II_{\boldsymbol{E}}}(f_0+f_1+f_2)\doteq(\vartheta-\delta_1)\cdot\frac{1}{2}\mu(\alpha_{03}+\alpha_{13}+\alpha_{-13}),$$
$$=(\vartheta-\delta_1)\cdot\tfrac{1}{2}\mu\,\alpha_4, \tag{11.21}$$

and

$$\delta_1\frac{\partial}{\partial I_{\boldsymbol{E}}}(f_1+2f_2)\doteq\delta_1\frac{\partial}{\partial I_{\boldsymbol{E}}}(h_1-h_{-1}),$$
$$\doteq\mu(\alpha_{11}-\alpha_{-11})\,\delta_1=\tfrac{1}{2}\mu\,\alpha_5\,\delta_1. \tag{11.22}$$

Putting (11.20), (11.21), and (11.22) into (11.18) yields

$$\frac{B}{\mu}\doteq\frac{1}{2}\alpha_1+1+\left(\alpha_3+\frac{1}{2}\alpha_4+\frac{1}{2}\alpha_5\right)\vartheta+\left(-\frac{1}{2}\alpha_4+\frac{1}{2}\alpha_5+\alpha_6\right)\delta_1. \tag{11.23}$$

By substituting (11.15) and (11.23) into (11.13) and then using (11.11) we obtain the definitive result for the *second-order speeds of longitudinal principal waves:*

$$\frac{\varrho_0 U_{11}^2}{\mu}\doteq\frac{\lambda}{\mu}+2+\left(\frac{\lambda}{\mu}+2+2\alpha_3+\alpha_4+\alpha_5\right)\vartheta+\left(\frac{2\lambda}{\mu}+4-\alpha_4+\alpha_5+2\alpha_6\right)\delta_1. \tag{11.24}$$

The formulae (11.12) and (11.24), giving the speeds of propagation explicitly and generally for the second-order theory, include and generalize a long series of particular results of this kind.

L. Brillouin [1925, Eqs. (76), (77)] [1938, Eqs. (XI.125), (XI.126)] was the first to calculate the second-order wave speeds for a general hyperelastic material subject to uniform hydrostatic pressure. To reduce our results to forms corresponding to his, put $\delta_1=\delta_2=\delta_3\equiv e$. Then $U_{11}^2=U_{22}^2=U_{33}^2\equiv U_{\|}^2$, and $U_{12}^2=U_{21}^2=\cdots=U_{32}^2\equiv U_{\perp}^2$, and from (11.12) and (11.24) we see that[3]

$$\frac{\varrho_0}{\mu}\frac{dU_{\|}^2}{de}=\frac{5\lambda}{\mu}+10+6\alpha_3+2\alpha_4+4\alpha_5+2\alpha_6,$$
$$\frac{\varrho_0}{\mu}\frac{dU_{\perp}^2}{de}=5+\frac{3}{2}\alpha_5+\alpha_6. \tag{11.25}$$

While L. Brillouin [1925, §11] had calculated the second-order speeds of principal waves in a medium subject to uniaxial tension, apparently Hughes & Kelly [1953, 2, Eqs. (12)] were the first to obtain a sufficient number of independent relations to determine the second-order elasticities from wave speeds in hyperelastic materials. Their method of linearization is not altogether clear to me, and I have not attempted to check their results against mine. Toupin & Bernstein [1961, 1, Eqs. (5.12)] obtained formulae which are easier to assess. They considered a biaxial tension. To obtain results corresponding to theirs, put $U_{11}^2=v_{\|}^2$, $U_{12}^2=v_m^2$, $U_{13}^2=v_n^2$, $\delta_1=\delta_3=-\sigma\delta_2=-\sigma e$,

[3] To get Brillouin's results, which presuppose a hyperelastic material, put

$$\mu\,\alpha_3=-\lambda+8A+24B,$$
$$\mu\,\alpha_4=-16A,$$
$$\mu\,\alpha_5=2(\lambda-\mu+8A),$$
$$\mu\,\alpha_6=4(\mu+6C). \tag{F}$$

where σ is to be held constant as $e\to 0$. Then from (11.12) and (11.24) we see that[4]

$$\begin{aligned}
\frac{\varrho_0}{\mu}\frac{dv_{\|}^2}{de} &= (1-2\sigma)\left(\frac{\lambda}{\mu}+2+2\alpha_3+\alpha_4+\alpha_5\right)-\sigma\left(\frac{2\lambda}{\mu}+4-\alpha_4+\alpha_5+2\alpha_6\right),\\
\frac{\varrho_0}{\mu}\frac{dv_m^2}{de} &= 1+\frac{1}{2}(1-2\sigma)\,\alpha_5+\frac{1}{2}(1-\sigma)\,\alpha_6,\\
\frac{\varrho_0}{\mu}\frac{dv_n^2}{de} &= 1+\frac{1}{2}(1-2\sigma)\,\alpha_5-\sigma\,\alpha_6.
\end{aligned}\tag{11.26}$$

TOUPIN & BERNSTEIN have developed alternative forms of these results, suitable for immediate comparison with experimental data.

Let us now suppose that the 9 squared wave speeds U_{ab}^2 are known, possibly from experimental data, as linear functions of the extensions δ_c. We shall find necessary and sufficient conditions that these linear expressions be compatible with second-order elasticity for some value of the coefficients α_Γ. First, the general conditions (9.1) and (9.2) for the transverse waves may be replaced by the following second-order approximations:

$$\begin{gathered}
\frac{U_{12}^2-U_{21}^2}{\delta_1-\delta_2}=\frac{U_{23}^2-U_{32}^2}{\delta_2-\delta_3}=\frac{U_{31}^2-U_{13}^2}{\delta_1-\delta_3}=2U_\perp^2,\\
U_{23}^2=2(\delta_2-\delta_1)\,U_\perp^2+\frac{1}{2}(U_{12}^2+U_{13}^2)+\frac{1}{2}\,\frac{U_{12}^2-U_{13}^2}{\delta_2-\delta_3}(\delta_2+\delta_3-2\delta_1),
\end{gathered}\tag{11.27}$$

where $U_\perp^2$ is the squared speed of transverse waves according to the linearized theory. Alternatively, these conditions may easily be derived directly from (11.12). When they are satisfied, unique values for α_5 and α_6 are determined from the squared speeds of the two kinds of transverse waves corresponding to any one principal axis. *E.g.*,

$$\begin{aligned}
\frac{1}{2}\alpha_6 &= \frac{U_{12}^2-U_{13}^2}{(\delta_2-\delta_3)\,U_\perp^2},\\
\frac{1}{2}\alpha_5 &= \frac{1}{\vartheta}\left[-2\delta_1-\frac{1}{4}(2\vartheta-\delta_3)\,\alpha_6+\frac{U_{12}^2+U_{13}^2}{2U_\perp^2}-1\right]-1.
\end{aligned}\tag{11.28}$$

For the longitudinal waves it is easier to procede directly from (11.24). In this way one sees that the three differential conditions (9.7) are equivalent in second-order approximation to a single algebraic one:

$$(U_{11}^2-U_{\|}^2)(\delta_2-\delta_3)+(U_{22}^2-U_{\|}^2)(\delta_3-\delta_1)+(U_{33}^2-U_{\|}^2)(\delta_1-\delta_2)=0,\tag{11.29}$$

where $U_{\|}^2$ is the squared speed of longitudinal waves according to the linearized theory. If this condition is satisfied, and if α_5 and α_6 are previously specified by (11.28) or otherwise, unique values for α_3 and α_4 are determined from the squared speeds of any two kinds of longitudinal waves. *E.g.*,

$$\begin{aligned}
\alpha_3 &= \frac{(U_{11}^2-U_{\|}^2)\,\delta_2-(U_{22}^2-U_{\|}^2)\,\delta_1}{2\nu(\delta_2-\delta_1)\,U_\perp^2}+\frac{U_{22}^2-U_{11}^2}{2(\delta_2-\delta_1)\,U_\perp^2}-\frac{3}{2}\left(2+\frac{\lambda}{\mu}\right)-\alpha_5-\alpha_6,\\
\alpha_4 &= -\frac{U_{22}^2-U_{11}^2}{(\delta_2-\delta_1)\,U_\perp^2}+2\left(2+\frac{\lambda}{\mu}\right)+\alpha_5+\alpha_6.
\end{aligned}\tag{11.30}$$

[4] To get TOUPIN & BERNSTEIN's results, which presuppose a hyperelastic material, put

$$\begin{aligned}
\mu\,\alpha_3 &= -\lambda+\nu_2+\tfrac{1}{2}\nu_1,\\
\mu\,\alpha_4 &= -2\nu_2,\\
\mu\,\alpha_5 &= 2(\lambda-\mu+\nu_2),\\
\mu\,\alpha_6 &= 4(\mu+\nu_3).
\end{aligned}\tag{G}$$

In any theory in which the squared speeds emerge as linear functions of the extensions, no satisfactory conditions for the speeds to be real can result, for no linear function is positive-definite. Since the speeds cannot be non-constant linear functions of the extensions according to any exact stress-strain relation, this result does not indicate any defect of the general theory. The second-order theory is only an approximation; the resulting formulae are expected to be valid only for small extensions; and for sufficiently small extensions, the second-order wave speeds are always real if the first-order ones are. Thus no restriction on the coefficients can be found by considering the second-order wave speeds. Restrictions may result by applying the various conditions of §10 to the third-order theory of elasticity, but I will not enter this forbidding avenue of research.

The formulae and analysis of this section when compared to those preceding furnish an example of the frequent observation that an exact and general theory is often simpler and easier to understand than an approximate one or the result of a perturbation process. It is possible, of course, but less enlightening to calculate the second-order wave speeds directly from the second-order stress-strain relations, given as Eq. (E) in the footnote on p. 283.

12. Hyperelastic materials

When there exists a strain energy $\Sigma(x^m{}_{,\beta}, \boldsymbol{d}_A)$, the function $\mathfrak{h}_k^\alpha$ in (2.3) assumes the form [CFT, Eq. (303.5)]

$$\mathfrak{h}_k^\alpha = \frac{\partial \Sigma}{\partial x^k{}_{,\alpha}}. \tag{12.1}$$

From (2.5) we see that then

$$A_{k\,m}^{\;\;\alpha\beta} = A_{m\,k}^{\;\;\beta\alpha}. \tag{12.2}$$

Now by (2.8),

$$Q_{k\,m}(\boldsymbol{n}) - Q_{m\,k}(\boldsymbol{n}) = [A_{k\,m}^{\;\;(\alpha\beta)} - A_{m\,k}^{\;\;(\alpha\beta)}]\, x^p{}_{,\alpha}\, x^q{}_{,\beta}\, n_p\, n_q. \tag{12.3}$$

Hence (12.2) implies that $Q_{k\,m} = Q_{m\,k}$; thus[1] *the acoustical tensor of a hyperelastic material is symmetric; consequently all acoustic axes are real, and for each direction of propagation there exists at least one mutually orthogonal set of acoustic axes*[2].

In the hyperelastic case, accordingly, much more specific results hold. When we apply the principle of material indifference, we may express all results in

[1] The condition is sufficient, not necessary. In order that $Q_{k\,m}(\boldsymbol{n}) = Q_{m\,k}(\boldsymbol{n})$ for every $\boldsymbol{n}$, it is necessary that

$$A_{k\,m}^{\;\;\alpha\beta} - A_{m\,k}^{\;\;\beta\alpha} = L_{k\,m}^{\;\;\alpha\beta}, \tag{H}$$

where

$$L_{k\,m}^{\;\;\alpha\beta} = L_{m\,k}^{\;\;\alpha\beta} = -L_{k\,m}^{\;\;\beta\alpha}. \tag{I}$$

In view of the theorem of BERNSTEIN given below, in order to construct a non-trivial example satisfying Eqs. (H) and (I) it is necessary to consider materials violating the principle of material indifference. Since no linear constitutive equation can satisfy that principle, take $\mathfrak{h}_k^\alpha = K_k{}^\alpha{}_m{}^\beta\, x^m{}_{,\beta}$, where $K_k{}^\alpha{}_m{}^\beta = \text{const}$. Then $A_{k\,m}^{\;\;\alpha\beta} = K_k{}^\alpha{}_m{}^\beta$, and solutions of Eqs. (H) and (I) are many.

[2] This theorem is due to HADAMARD [1901] [1903, 1, p. 268], generalizing earlier work of CHRISTOFFEL and HUGONIOT.

HADAMARD's theorem was questioned by DUHEM [1903] [1906. Ch. I, §§ II, IV]; while I have not attempted to follow the long analysis of DUHEM, I suspect that only some misunderstanding of notation is involved.

terms of the tensors B^{kmpq} and C^{kmpq} of §6. From (6.7) and (12.2), then,

$$B^{kmpq} = B^{pqkm}; \tag{12.4}$$

consequently, by (6.9),

$$C^{kmpq} = C^{pqkm}. \tag{12.5}$$

Eqs. (12.2), (12.4), and (12.5) are alternative necessary and sufficient conditions that an elastic material be hyperelastic.

BERNSTEIN[2a] has pointed out that when the principle of material indifference is adopted, a definite and simple converse to the foregoing theorem comes into force: *If for every direction of propagation the acoustical tensor of an elastic material is symmetric, then the material is hyperelastic.* For by (6.11) we see that $Q^{km}(\boldsymbol{n}) = Q^{mk}(\boldsymbol{n})$ if and only if

$$(B^{kpmq} - B^{mpkq})\, n_p n_q = 0. \tag{12.5 A}$$

For this condition to hold for every $\boldsymbol{n}$, it is necessary and sufficient that

$$B^{kpmq} - B^{mpkq} + B^{kqmp} - B^{mqkp} = 0. \tag{12.5 B}$$

Hence

$$B^{pkqm} - B^{qkpm} + B^{pmqk} - B^{qmpk} = 0. \tag{12.5 C}$$

Adding these two equations yields

$$B^{kpmq} + B^{pkqm} - B^{mqkp} - B^{qmpk} = B^{mpkq} - B^{pmqk} + B^{qkpm} - B^{kqmp}. \tag{12.5 D}$$

By four uses of the identity $(6.10)_3$ we see that the right-hand side vanishes, while the left-hand side reduces to $2(B^{kpmq} - B^{kqmp})$. Hence (12.4) holds, showing that the material is hyperelastic.

It would be useful to express the theorem and its converse directly in terms of acoustical data, but this does not seem to be possible. Rephrasing a theorem of KELVIN & TAIT on matrices [CFT, App. §37], we see that if the squared speeds are non-negative and if an orthogonal triad of acoustical axes exists, then the acoustical tensor is symmetric. From this fact and from BERNSTEIN's theorem we may conclude that *if, in an elastic material, there exist three orthogonal acoustical axes for every direction of propagation, and if the corresponding speeds of propagation are real or purely imaginary, the material is hyperelastic.* This result fails to be a true physical converse to HADAMARD's theorem because purely imaginary speeds of propagation do not seem to be distinguishable from complex speeds, since neither correspond to real singular surfaces, while both lead to amplified infinitesimal oscillations. However, it shows that the existence of orthogonal axes with real speeds of propagation for every direction is *sufficient* for an elastic material to be hyperelastic. Of course it is possible that in an elastic material which is not hyperelastic, the acoustical tensor may be symmetric for some particular directions of propagation. For example, it follows at once from (7.8) that the acoustical tensor for any principal direction of stress in an isotropic material is always symmetric, whether or not there be a stored energy.

[*Note added in proof, September 9, 1961.* The case last mentioned in small type is truly exceptional, for I have just discovered that *if in an isotropic elastic material the acoustical tensor $\boldsymbol{Q}(\boldsymbol{n})$ is symmetric for one single direction $\boldsymbol{n}$ not lying in the plane of two principal directions, then the material is hyperelastic.* To prove this result, we begin by writing out the components of $T^{km}{}_{pq}$ in a principal co-

[2a] Letters dated July 28 and August 8, 1961. The paragraph beginning here and the following text in small type were added in proof, September 7, 1961.

ordinate system. Using (7.12), we see from (7.5) that

$$T^{11}{}_{11} = \frac{\partial t_1}{\partial v_1^2}, \qquad T^{11}{}_{22} = \frac{\partial t_1}{\partial v_2^2}, \ldots$$
$$T^{12}{}_{12} = \frac{t_1 - t_2}{2(v_1^2 - v_2^2)}, \ldots, \tag{12.5 E}$$

while the other types of components of $T^{km}{}_{pq}$ vanish. Let the components of $\boldsymbol{n}$ be $\cos\vartheta_1$, $\cos\vartheta_2$, $\cos\vartheta_3$; then from (7.7) we find that

$$Q_2^1(\boldsymbol{n}) = \frac{\tilde{\varrho}}{\varrho}\left[\frac{\partial t_1}{\partial \log v_2} + \frac{t_1 - t_2}{v_1^2 - v_2^2} v_1^2\right] \cos\vartheta_1 \cos\vartheta_2, \tag{12.5 F}$$

while Q_1^2 is obtained by interchanging 1 and 2 in this formula. For a principal wave, $\cos\vartheta_1 \cos\vartheta_2 = 0$, so that $Q_2^1 = Q_1^2 = 0$, as we have shown in §7 by a different argument. If, however, $\cos\vartheta_1 \cos\vartheta_2 \neq 0$, then $Q_2^1 = Q_1^2$ if and only if

$$\frac{\partial t_2}{\partial \log v_1} - \frac{\partial t_1}{\partial \log v_2} = t_2 - t_1. \tag{12.5 G}$$

In an earlier work [1952, Eq. $(41.11)_2$] I have shown that this equation and the similar ones obtained from it by permutations of indices are necessary and sufficient conditions for the existence of a stored energy. Q.E.D.

If $\cos\vartheta_3 = 0$ but $\cos\vartheta_1 \cos\vartheta_2 \neq 0$, so that the wave is not a principal wave, but $\boldsymbol{n}$ lies in the plane of the principal directions corresponding to the principal stretches v_1 and v_2, then $Q_3^2 = Q_2^3 = Q_1^3 = Q_3^1 = 0$, but the single condition (12.5 G) is necessary and sufficient that $Q_2^1 = Q_1^2$. It is thus possible for special isotropic elastic but not hyperelastic materials to have symmetric acoustical tensors for all directions of propagation lying in one and only one of the three planes defined by the pairs of principal axes.

Therefore, there are three possible kinds of isotropic materials as far as acoustical properties are concerned:

1. Those for which $\boldsymbol{Q}(\boldsymbol{n})$ is symmetric for the principal waves only.

2. Those for which $\boldsymbol{Q}(\boldsymbol{n})$ is symmetric for the principal waves and also for any $\boldsymbol{n}$ perpendicular to one and only one of the principal axes.

3. Those for which $\boldsymbol{Q}(\boldsymbol{n})$ is symmetric for all $\boldsymbol{n}$, these last constituting the hyperelastic materials.]

Three further definite advances over the results of the more general theory may now be made. First, if we set

$$P^{pq} \equiv B_k^{pkq}, \tag{12.6}$$

from (12.4) it follows that $\boldsymbol{P}$ is symmetric. Also, if for a unit vector $\boldsymbol{n}$ we set

$$m(\boldsymbol{n}) \equiv P^{pq} n_p n_q; \tag{12.7}$$

from (6.11) and (2.9) we see that

$$m(\boldsymbol{n}) = \frac{\varrho}{\tilde{\varrho}} Q_k^k(\boldsymbol{n}) = \varrho \sum_{a=1}^{3} U_a^2, \tag{12.8}$$

where $\tilde{\varrho}\, U_a^2$ is the ath proper number of $\boldsymbol{Q}(\boldsymbol{n})$. Since $\boldsymbol{P}$ is real and symmetric, there are at least three mutually orthogonal directions $\boldsymbol{n}$ in which $m(\boldsymbol{n})$, as given by (12.7), is extreme. By (12.8), then, *in a hyperelastic material there exist at each point at least three mutually orthogonal directions* $\boldsymbol{n}_a$ *such that the sum of the squares of the speeds of propagation of the possible waves travelling in the* *direction $\boldsymbol{n}_a$ is an extremum*[3]. In the foregoing statement, the possibility that some wave speeds are purely imaginary is not excluded.

Second, as is clear from (9.2), *the S-E condition is necessary and sufficient that all squared wave speeds be positive.* In particular, if the *S-E* condition is satisfied, there can be no material singular surfaces.

Third, *the C-N condition ensures that at least one wave may travel in the direction of $\boldsymbol{n}$, for any $\boldsymbol{n}$.* This result was discovered by NOLL[4] and was proved as follows by ERICKSEN[5]. The proper numbers of $\boldsymbol{Q}(\boldsymbol{n})$ are the extrema of the bounded continuous function

$$F(\boldsymbol{l}) \equiv Q^{km}(\boldsymbol{n})\, l_k\, l_m, \tag{12.9}$$

where $\boldsymbol{l}$ is a unit vector. By (10.5), $F(\boldsymbol{n}) > 0$. Hence at least one extreme of $F(\boldsymbol{l})$ must be positive; that is, $\boldsymbol{Q}(\boldsymbol{n})$ has at least one positive proper number, so that the corresponding wave speed is real.

For isotropic materials it is easy to show[6] that the response coefficients f_Γ in (7.1) have the following special forms:

$$\frac{1}{2} f_0 = \sqrt{III}\,\frac{\partial\Sigma}{\partial III}, \qquad \frac{1}{2}\frac{\varrho_0}{\varrho} f_1 = \frac{\partial\Sigma}{\partial I} + I\,\frac{\partial\Sigma}{\partial II}, \qquad \frac{1}{2}\frac{\varrho_0}{\varrho} f_2 = -\frac{\partial\Sigma}{\partial II}, \tag{12.10}$$

where $\Sigma(I, II, III)$ is the strain energy. Hence

$$2\frac{\partial\Sigma}{\partial I} = \frac{\tilde\varrho}{\varrho}(f_1 + I f_2), \qquad 2\frac{\partial\Sigma}{\partial II} = -\frac{\tilde\varrho}{\varrho} f_2, \qquad 2\frac{\partial\Sigma}{\partial III} = \frac{1}{\sqrt{III}} f_0. \tag{12.11}$$

Accordingly, conditions of integrability for Σ are

$$\begin{gathered}
\frac{\partial}{\partial I}\left(\frac{f_2}{\varrho}\right) + \frac{\partial}{\partial II}\left(\frac{f_1}{\varrho} + I\frac{f_2}{\varrho}\right) = 0,\\
\frac{1}{\tilde\varrho\sqrt{III}}\frac{\partial f_0}{\partial I} - \frac{\partial}{\partial III}\left(\frac{f_1}{\varrho} + I\frac{f_2}{\varrho}\right) = 0,\\
\frac{1}{\tilde\varrho\sqrt{III}}\frac{\partial f_0}{\partial II} + \frac{\partial}{\partial III}\left(\frac{f_2}{\varrho}\right) = 0.
\end{gathered} \tag{12.12}$$

By use of (9.3) and (9.6) we may express these conditions directly in terms of the wave speeds:

$$\begin{gathered}
\frac{\partial}{\partial I}\left[\frac{U_{12}^2 - U_{13}^2}{v_1^2(v_2^2 - v_3^2)}\right] + \frac{\partial}{\partial II}\left[\frac{v_2^2 U_{12}^2 - v_3^2 U_{13}^2}{v_1^2(v_2^2 - v_3^2)}\right] = 0,\\
\frac{1}{\tilde\varrho\sqrt{III}} B_1 - \frac{\partial}{\partial III}\left[\frac{v_2^2 U_{12}^2 - v_3^2 U_{13}^2}{v_1^2(v_2^2 - v_3^2)}\right] = 0,\\
\frac{1}{\tilde\varrho\sqrt{III}} B_2 + \frac{\partial}{\partial III}\left[\frac{U_{12}^2 - U_{13}^2}{v_1^2(v_2^2 - v_3^2)}\right] = 0.
\end{gathered} \tag{12.13}$$

We presume that all the conditions of compatibility obtained in §9 are satisfied; then, in particular, B_1 and B_2 and the two expressions in brackets in (11.13)

[3] This result generalizes slightly a theorem of PASTORI [1949, §6]. It is obvious that $\varrho\sum_{a=1}^{3} U_a^2(\boldsymbol{n})$, being a bounded continuous real function of $\boldsymbol{n}$, has a maximum and a minimum unless it is constant. It is the existence of the third extremum that does not seem evident *a priori*.

[4] Letter dated September 13, 1960.

[5] Letter dated June 16, 1961. In a letter of July 14, 1961, NOLL informed me that this proof is the same as his and disclosed a sharper inequality for $F(\boldsymbol{l})$.

[6] The results used here are equivalent to those given by TRUESDELL [1952, Eq. (41.5)] and may be derived by a similar procedure.

are symmetric functions of the v_a and hence are equal to functions of I, II, III. By use of the conditions of compatibility, moreover, A_1, A_2, and A_3, as defined by (9.4), are expressible in terms of U_{12}^2, U_{13}^2, U_{11}^2, U_{22}^2, U_{33}^2, their first derivatives, and v_1, v_2, v_3; thus B_1 and B_2, as defined by $(9.6)_{1,2}$, are functions of these same quantities. The results obtained may be summarized as follows: *for an isotropic elastic material, let the three possible squared wave-speeds corresponding to one principal axis of stress and strain as well as the remaining two longitudinal wave-speeds be given as functions of the principal stretches; then the material is hyperelastic if and only if the three differential conditions* (12.13) *are satisfied.*

The conditions (12.13), while expressed directly in terms of the speeds of principal waves, are elaborate in form and difficult to interpret. BERNSTEIN'S condition, namely, that $\boldsymbol{Q}(\boldsymbol{n})$ shall be symmetric for every $\boldsymbol{n}$, is simple but does not seem to lend itself to interpretation in terms of acoustical data alone. It assures us that an elastic material in which three waves of mutually orthogonal amplitudes may propagate in any direction must have a stored energy, but this is not a characterizing condition. In respect to propagation of waves, there seems to be no especially simple property by which the existence or non-existence of a stored energy manifests itself without fail.

In the second-order theory, however, a simple identity, suitable for experimental test, replaces the elaborate general conditions (12.13). Either from (12.13), or from the formulae (E), (F), or (G) given in the footnotes on pp. 283, 285, and 286, respectively, it is easy to see that for the existence of a stored energy in the second-order theory of isotropic materials it is necessary and sufficient that the 6 first-order and second-order elasticities be reduced to 5 by the following equation:

$$\alpha_4 + \alpha_5 = 2\left(\frac{\lambda}{\mu} - 1\right). \tag{12.14}$$

This condition may be expressed directly in terms of the wave speeds by means of (11.28) and (11.29), but the resulting identity is complicated. An equivalent but simpler formula may be obtained by differentiation of (11.12) and (11.24). Since

$$\begin{aligned} \frac{\varrho_0}{\mu}\,\frac{\partial U_{12}^2}{\partial \delta_1} &= 3 + \frac{1}{2}\alpha_5 + \frac{1}{2}\alpha_6, \\ \frac{\varrho_0}{\mu}\left(\frac{\partial}{\partial \delta_1} - \frac{\partial}{\partial \delta_2}\right) U_{11}^2 &= \frac{2\lambda}{\mu} + 4 - \alpha_4 + \alpha_5 + 2\alpha_6, \end{aligned} \tag{12.15}$$

we see that (12.14) is equivalent to the following ***universal criterion for a stored energy*** in the second-order theory:

$$\left(\frac{\partial}{\partial \delta_1} - \frac{\partial}{\partial \delta_2}\right) U_{11}^2 - 4\,\frac{\partial U_{12}^2}{\partial \delta_1} + 6\,U_{\perp}^2 = 0. \tag{12.16}$$

This condition affords the first opportunity, so far as I know, for a clear-cut and unequivocal experimental test of the existence of a stored energy in an isotropic material[7]. It is different from the previously derived conditions of

[7] In the linearized theory, every isotropic elastic material is hyperelastic. In the finite theory, this is not so; while practically all previous researches on the finite theory concern hyperelastic materials, I know of no case in which the existence of a stored energy is necessary for the results obtained. In fact, many of the solutions derived by the British school could be expressed more simply in terms of the coefficients h_Γ, independently of whether there be a stored energy.

compatibility in that *it connects a longitudinal-wave speed with a transverse-wave speed.* There are in fact 6 conditions of this kind, but, on the presumption that the general identities (11.27) and (11.29) are satisfied, any of them implies the rest. In this sense (12.16) is *necessary and sufficient* for the existence of a second-order stored energy. If the partial derivatives are evaluated in the natural state, (12.16) is also a necessary but not sufficient condition for the existence of a stored energy according to the general theory.

In order to test (12.16) properly it is necessary to let δ_1, δ_2 and δ_3 vary *independently*. It is impossible, for example, to find a relation corresponding to (12.16) by considering a one-parameter family of deformations such as that used by TOUPIN & BERNSTEIN and explained in § 11. It may be preferable, however, to reduce the number of independent variables from three to two. For example, if we employ only deformations such that $\delta_2 = \delta_3$, then it is easy to show that instead of (12.16) we have the alternative necessary and sufficient condition

$$\left(\frac{\partial}{\partial \delta_1} - \frac{1}{2}\frac{\partial}{\partial \delta_2}\right) U_{11}^2 - 4\frac{\partial U_{12}^2}{\partial \delta_1} + 6U_{\perp}^2 = 0, \tag{12.17}$$

but δ_1 and δ_2 must be allowed to vary independently.

13. Duhem's theorems on the effect of temperature variation

In general the stress-strain relation (2.3) involves an additional scalar parameter such as the temperature, ϑ, or the specific entropy, η:

$$T_k^\alpha = \hat{\mathfrak{h}}_k^\alpha(x^m{}_{,\beta}, \boldsymbol{d}_A, \eta) = \bar{\mathfrak{h}}_k^\alpha(x^m{}_{,\beta}, \boldsymbol{d}_A, \vartheta), \tag{13.1}$$

where $\hat{\mathfrak{h}}$ and $\bar{\mathfrak{h}}$ are different functions. The results obtained so far are valid when $\eta = \text{const.}$ or when $\vartheta = \text{const.}$, but the tensors $\boldsymbol{A}$, defined by (2.5), and the acoustical tensors, $\boldsymbol{Q}$, defined by (2.7), are different in the two cases and may be distinguished by the notations $\hat{\boldsymbol{A}}, \bar{\boldsymbol{A}}, \hat{\boldsymbol{Q}}, \bar{\boldsymbol{Q}}$.

As far as I can learn, the only treatment of more general conditions, when temperature and entropy may vary within the body, is that given by DUHEM [1903, 2] [1906, Ch. I, §2]. I have despaired at following his long calculations, made more difficult from an unfortunate choice of variables, but here I present my own treatment of the same problems, leading to results which seem to be equivalent to his.

First, if we substitute (13.1) into (2.4), we obtain equations of motion of the following forms:

$$\begin{aligned}
\hat{A}_{km}{}^{(\alpha\beta)} x^m{}_{;\alpha\beta} + \frac{\partial \hat{\mathfrak{h}}_k^\alpha}{\partial d_A^m} d^m_{A;\alpha} + \frac{\partial \hat{\mathfrak{h}}_k^\alpha}{\partial \eta} \eta_{,m} x^m{}_{,\alpha} + \tilde{\varrho} f_k &= \tilde{\varrho} \ddot{x}_k, \\
\bar{A}_{km}{}^{(\alpha\beta)} x^m{}_{;\alpha\beta} + \frac{\partial \bar{\mathfrak{h}}_k^\alpha}{\partial d_A^m} d^m_{A;\alpha} + \frac{\partial \bar{\mathfrak{h}}_k^\alpha}{\partial \vartheta} \vartheta_{,m} x^m{}_{,\alpha} + \tilde{\varrho} f_k &= \tilde{\varrho} \ddot{x}_k.
\end{aligned} \tag{13.2}$$

Taking the jump of these equations at a singular surface leads to propagation conditions of the form (2.7) not only if $\eta = \text{const.}$ or if $\vartheta = \text{const.}$ but also under the weaker assumptions

$$[\eta_{,m}] = 0, \quad [\vartheta_{,m}] = 0, \tag{13.3}$$

respectively. We now consider physical circumstances such as to render valid one or the other of the relations (13.3).

First, we review a classical argument, given in effect by FOURIER [*cf.* CFT, §296], which shows that if $[\vartheta]=0$ and if FOURIER's law of heat conduction is assumed, then $(13.3)_2$ follows. For by MAXWELL's theorem [CFT, Eq. (175.8)], the condition $[\vartheta]=0$ implies that $[\vartheta_{,k}]=B\,n_k$. FOURIER's law of heat conduction asserts that the flux of energy, $\boldsymbol{h}$, depends linearly on $\vartheta_{,k}$, that is, $h_m=K_m{}^p\vartheta_{,p}$, where $\boldsymbol{K}$ is positive-definite. Hence $[h_m]=BK_m{}^p n_p$, where $\boldsymbol{K}$ is assumed continuous, and consequently $\boldsymbol{n}\cdot[\boldsymbol{h}]=BK_m{}^p n_p n^m$. But the principle of conservation of energy requires that $\boldsymbol{n}\cdot[\boldsymbol{h}]=0$, provided the internal energy be continuous. Hence $B=0$, whence $(13.3)_2$ follows.

The foregoing argument has nothing to do with elasticity, but we may summarize its implication in an elastic context by the following ***theorem on elastic conductors:*** *In an elastic material subject to* FOURIER'S *law of heat conduction, the laws of propagation for waves of second order across which the temperature, internal energy, and heat conductivity are continuous are the same as those of homothermal waves.* Similarly, if we consider a wave of third order across which $\dot{\varepsilon}$, $\vartheta_{,k}$, $K_m{}^p$, and $K_m{}^p{}_{,q}$ are assumed continuous, from the differential form of the equation of energy we conclude that $[\vartheta_{,km}]=0$, whence it follows that the laws of propagation are again those of homothermal waves. Parallel assumptions lead to a parallel theorem for waves of higher order.

Second, assume that there is no conduction of heat at all: $\boldsymbol{h}=0$. Then[1] $\dot{\eta}=0$, and consequently $[\dot{\eta}]=0$. If it is assumed that $[\eta]=0$, then by HADAMARD's conditions [CFT, Eq. (180.5)] we infer that $(13.3)_1$ holds. Thus we have the following ***theorem on elastic non-conductors***[2]***:*** *in an elastic material that does not conduct heat, the laws of propagation of acceleration waves are the same as for the homentropic case.* Parallel assumptions lead to a parallel theorem for waves of higher order.

We may relate the tensors $\hat{\boldsymbol{A}}$ and $\bar{\boldsymbol{A}}$ if we assume an equation of state for the temperature:

$$\vartheta=\hat{\vartheta}(x^m{}_{,\beta},\boldsymbol{d}_\Lambda,\eta)\,. \tag{13.4}$$

Then from (13.1) and (2.5) we see that

$$\hat{A}_{km}{}^{\alpha\beta}=\bar{A}_{km}{}^{\alpha\beta}+\frac{\partial\bar{\mathfrak{h}}_k^\alpha}{\partial\vartheta}\,\frac{\partial\hat{\vartheta}}{\partial x^m{}_{,\beta}}\,. \tag{13.5}$$

While this result is not illuminating in general, for hyperelastic materials it can be put into a form having immediate interpretation. In the hyperelastic case there is an equation of state for the internal energy,

$$\varepsilon=\hat{\varepsilon}\,(x^p{}_{,\gamma},\boldsymbol{d}_\Lambda,\eta)\,, \tag{13.6}$$

while

$$T_k^\alpha=\hat{\mathfrak{h}}_k^\alpha=\tilde{\varrho}\,\frac{\partial\hat{\varepsilon}}{\partial x^k{}_{,\alpha}}\,,\qquad \vartheta=\hat{\vartheta}=\frac{\partial\hat{\varepsilon}}{\partial\eta} \tag{13.7}$$

[*cf.* CFT, §256A]. Then

$$\frac{\partial\hat{\vartheta}}{\partial x^m{}_{,\beta}}=\frac{\partial^2\hat{\varepsilon}}{\partial x^m{}_{,\beta}\,\partial\eta}=\frac{1}{\tilde{\varrho}}\,\frac{\partial\hat{\mathfrak{h}}_k^\alpha}{\partial\eta}\,. \tag{13.8}$$

---

[1] For hyperelastic materials, this may be proved [*cf.* CFT, §256A]; for more general elastic materials, it is offered as a hypothesis.

[2] The special case of this theorem for gas dynamics is familiar: The speed of sound for isentropic flow is the same as for a homentropic flow at the same thermodynamic state.

But by substituting $(13.7)_3$ into $(13.1)_2$ and differentiating with respect to η we see that

$$\frac{\partial \hat{\mathfrak{h}}_m^{\ \beta}}{\partial \eta} = \frac{\partial \bar{\mathfrak{h}}_m^{\ \beta}}{\partial \vartheta} \frac{\partial \hat{\vartheta}}{\partial \eta} = \frac{\vartheta}{\varkappa} \frac{\partial \bar{\mathfrak{h}}_m^{\ \beta}}{\partial \vartheta}, \tag{13.9}$$

where $\varkappa$ is the specific heat at constant deformation. Substituting (13.9) into (13.5) yields a fundamental relation between the homentropic and homothermal elasticities of any hyperelastic material:

$$\hat{A}_{k\,m}^{\ \ \alpha\beta} = \bar{A}_{k\,m}^{\ \ \alpha\beta} + \frac{\vartheta}{\varkappa\,\varrho} \frac{\partial \bar{\mathfrak{h}}_k^{\ \alpha}}{\partial \vartheta} \frac{\partial \bar{\mathfrak{h}}_m^{\ \beta}}{\partial \vartheta}. \tag{13.10}$$

Using (2.8), we derive DUHEM'S ***fundamental relation*** between the two acoustical tensors:

$$\hat{Q}_{k\,m} = \bar{Q}_{k\,m} + \frac{\vartheta}{\varkappa\,\varrho} P_k P_m \tag{13.11}$$

where

$$P_k \equiv \frac{\partial \bar{\mathfrak{h}}_k^{\ \alpha}}{\partial \vartheta} x^p_{\ ,\alpha} n_p. \tag{13.12}$$

Thus the properties of homentropic waves may be calculated directly from the stress-strain relations in which temperature is the thermodynamic parameter. Moreover, the tensor $P_k P_m$ is always non-negative definite, for any $\boldsymbol{n}$. Therefore, since $\vartheta/\varkappa \geqq 0$, it follows that if $\bar{\boldsymbol{Q}}(\boldsymbol{n})$ is positive-definite, so is $\hat{\boldsymbol{Q}}(\boldsymbol{n})$. Thus, as remarked by DUHEM, *if, for a given hyperelastic material subject to given deformation, three independent waves may propagate in the direction* $\boldsymbol{n}$ *if* $\vartheta =$ const., *then the same holds if* $\eta =$ const. In other words, if the homothermal wave speeds are real, so are the homentropic wave speeds, but the converse need not hold.

14. Remarks on other researches

In this paper I have not attempted to summarize everything known regarding waves in severely deformed elastic materials[1], but some topics should be listed for reference.

1. The connection between HADAMARD'S stability and wave propagation, already mentioned.

2. JOUGUET'S presentation of the general shock conditions in an elastic context and analysis of weak shocks [1920, 1—4] [1921].

3. L. BRILLOUIN'S elaborate study of radiation pressure and other second-order effects [1925] [1938, Ch. XI, §§ VII—XVI].

4. ERICKSEN'S analysis [1953, 1] of waves in isotropic incompressible hyperelastic materials. By combining the approach of ERICKSEN with that presented in §§ 2—6 of this paper, a complete theory of waves in general incompressible elastic materials is easily constructed. Such waves are always transverse [CFT, §191]. For isotropic materials the wave speeds are given by (7.10); the relation (7.15), the conditions of compatibility (9.1) and (9.2), and the determination of the stress-strain relations from the wave speeds by means of (9.3) remain valid but may be simplified a little since $v_1^2 v_2^2 v_3^2 = 1$. The second-order theory for a general isotropic incompressible material may be read off from ERICKSEN'S

[1] Such a summary will appear in the article by W. NOLL & C. TRUESDELL, *The Non-linear Field Theories of Mechanics*, FLÜGGE'S Handbuch der Physik, Vol. 8, part III, in preparation.

exact theory for a Mooney material, since such a material serves as a general second-order approximation. In addition to the 4 conditions of compatibility (11.27), there is 1 more:

$$\frac{U_{12}^2+U_{13}^2-2U_\perp^2}{\delta_1}-\frac{U_{12}^2-U_{13}^2}{\delta_2-\delta_3}=4U_\perp^2. \qquad (14.1)$$

Thus any one speed determines all possible speeds; furthermore, the single second-order elasticity is also determined by any one wave speed; in fact, it suffices to consider a one-parameter family of deformations. As is well known, every isotropic incompressible elastic material is also hyperelastic as far as the second-order theory is concerned.

5. TOUPIN & BERNSTEIN's conditions of compatibility for waves according to the linearized theory of anisotropic materials [1961, 1, §4]. These are easily generalized, in slightly modified form, to the case when there need be no stored energy.

I am happy to acknowledge the support of the U. S. National Science Foundation through a fellowship at Bologna for the period during which this paper was written. It is a pleasure to express my gratitude also to Messrs. BERNSTEIN, ERICKSEN, NOLL, and TOUPIN for their criticism and help.

Note added in proof, September 19, 1961. Shock relations in hyperelastic materials are derived and discussed qualitatively by J. DEWEY: Strong shocks and stress-strain relations in solids. Part I. Ballistic Research Laboratories Report No. 1074, Aberdeen Proving Grounds, Maryland, April, 1959. Plane infinitesimal oscillations in a biaxially stretched isotropic hyperelastic material that conducts heat according to thermoelastic constitutive equations,

$$\boldsymbol{\Sigma}=\boldsymbol{\Sigma}(\boldsymbol{B},\vartheta), \qquad \boldsymbol{h}=\boldsymbol{K}\cdot\nabla\vartheta,$$

where $\boldsymbol{K}$ is an isotropic function of $\boldsymbol{B}$ and $\nabla\vartheta$, have been studied by J.N. FLAVIN & A.E. GREEN: J. Mech. Phys. Solids **9**, 179—190 (1961). Except in limiting cases, the contents of these papers does not overlap the research presented above.

Not included in the list of earlier works cited by TOUPIN & BERNSTEIN are the following papers on special theories of waves in stressed materials: BIRCH, F.: The effect of pressure upon the elastic parameters of isotropic solids, according to MURNAGHAN'S theory of finite strain. J. Appl. Phys. **9**, 279—288 (1938). — BIOT, M. A.: The influence of initial stress on elastic waves. J. Appl. Phys. **11**, 522—530 (1940).

The somewhat perplexing conclusions of these authors regarding waves in bodies subject to great hydrostatic pressure are cleared and unified by the exact theory given in §8 of the foregoing paper.

References

1901 HADAMARD, J.: Sur la propagation des ondes. Bull. Soc. Math. France **29**, 50—60.

1903 1. HADAMARD, J.: Leçons sur la Propagation des Ondes et les Equations de l'Hydrodynamique (Lectures of 1898—1900). Paris: Hermann.

2. DUHEM, P.: Sur la propagation des ondes dans un milieu parfaitement élastique affecté de déformations finies. C. R. Acad. Sci. Paris **136**, 1379—1381.

1905 DUHEM, P.: Recherches sur l'élasticité. Troisième Partie. La stabilité des milieux élastiques. Ann. Ecole Norm. (3) **22**, 143—217.

1906 DUHEM, P.: Recherches sur l'élasticité. Quatrième Partie. Propriétés générales des ondes dans les milieux visqueux et non visqueux. Ann. Ecole Normale (3) **23**, 169—223.

1920 1. JOUGUET, E.: Sur les ondes de choc dans les corps solides. C. R. Acad. Sci. Paris **171**, 461—464.
2. JOUGUET, E.: Sur la celerité des ondes dans les corps solides. C. R. Acad. Sci. Paris **171**, 512—515.
3. JOUGUET, E.: Sur la variation d'entropie dans les ondes de choc des solides élastiques. C. R. Acad. Sci. Paris **171**, 789—791.
4. JOUGUET, E.: Application du principe de Carnot-Clausius aux ondes de choc des solides élastiques. C. R. Acad. Sci. Paris **171**, 904—907.
1921 JOUGUET, E.: Notes sur la théorie de l'élasticité. Ann. Toulouse (3) **12** (1920), 47—92.
1925 BRILLOUIN, L.: Sur les tensions de radiation. Annales de Physique (10) **4**, 528—586.
1938 BRILLOUIN, L.: Les Tenseurs en Mécanique et en Elasticité. Paris: Masson.
1942 FINZI, B.: Propagazione ondosa nei continui anisotropi. Rend. Ist. Lombardo Cl. Sci. Mat. Nat. **75**=(3) **6** (1941—1942), 630—640.
1943 TOLOTTI, C.: Deformazioni elastiche finite: onde ordinarie di discontinuità e caso tipico di solidi elastici isotropi. Rend. di Mat. e delle Applic. (5) **4**, 34—59.
1949 PASTORI, M.: Propagazione ondosa nei continui anisotropi e corrispondenti direzioni principali. Il Nuovo Cimento (9) **6**, 187—193.
1952 TRUESDELL, C.: The mechanical foundations of elasticity and fluid dynamics. J. Rational Mech. Anal. **1**, 125—300. Corrections and additions. *ibid.* **2**, 595—616 (1953).
1953 1. ERICKSEN, J. L.: On the propagation of waves in isotropic incompressible perfectly elastic materials. J. Rational Mech. Anal. **2**, 329—337.
2. HUGHES, D. S., & J. L. KELLY: Second-order elastic deformation of solids. Phys. Rev. (2) **92**, 1145—1149.
1954 BAKER, M., & J. L. ERICKSEN: Inequalities restricting the form of the stress-deformation relations for isotropic solids and Reiner-Rivlin fluids. J. Washington Acad. Sci. **44**, 33—35.
1956 TRUESDELL, C.: Das ungelöste Hauptproblem der endlichen Elastizitätstheorie. Z. angew. Math. Mech. **36**, 97—103.
1958 NOLL, W.: A mathematical theory of the mechanical behavior of continuous media. Arch. Rational Mech. Anal. **2** (1958/59), 197—226.
1959 1. COLEMAN, B. D., & W. NOLL: On the thermostatics of continuous media. Arch. Rational Mech. Anal. **4** (1959/60), 97—128.
2. RIVLIN, R. S.: The constitutive equations for certain classes of deformations. Arch. Rational Mech. Anal. **3**, 304—311.
1960 RIVLIN, R. S.: Constitutive equations for classes of deformations. Viscoelasticity: Phenomenological Aspects. New York: Academic Press 93—108.
1961 1. TOUPIN, R. A., & B. BERNSTEIN: Sound waves in deformed perfectly elastic materials; the acoustoelastic effect. J. Acoust. Soc. Amer. **33**, 216—225.
2. HAYES, M., & R. S. RIVLIN: Propagation of a plane wave in an isotropic elastic material subject to pure homogeneous deformation. Arch. Rational Mech. Anal. **8**, 15—22.
3. TOUPIN, R. A.: Some relations between waves, stability, uniqueness criteria, and restrictions on the form of the energy function in elasticity theory. Lecture delivered at a meeting in Newcastle-upon-Tyne.

The Johns Hopkins University
Baltimore, Maryland

(Received July 29, 1961)

MIX
Papier aus verantwortungsvollen Quellen
Paper from responsible sources
FSC® C105338

If you have any concerns about our products,
you can contact us on
ProductSafety@springernature.com

In case Publisher is established outside the EU,
the EU authorized representative is:
Springer Nature Customer Service Center GmbH
Europaplatz 3, 69115 Heidelberg, Germany

Printed by Libri Plureos GmbH
in Hamburg, Germany